Energy Management Systems for Optimal Operation of Electrical Micro/Nanogrids

Energy Management Systems for Optimal Operation of Electrical Micro/Nanogrids

Editor

Maria Carmela Di Piazza

MDPI • Basel • Beijing • Wuhan • Barcelona • Belgrade • Manchester • Tokyo • Cluj • Tianjin

Editor
Maria Carmela Di Piazza
Istituto di Ingegneria del Mare
(INM)
Italy

Editorial Office
MDPI
St. Alban-Anlage 66
4052 Basel, Switzerland

This is a reprint of articles from the Special Issue published online in the open access journal *Energies* (ISSN 1996-1073) (available at: https://www.mdpi.com/journal/energies/special_issues/electrical_micro_nanogrids).

For citation purposes, cite each article independently as indicated on the article page online and as indicated below:

LastName, A.A.; LastName, B.B.; LastName, C.C. Article Title. *Journal Name* **Year**, *Volume Number*, Page Range.

ISBN 978-3-0365-2558-7 (Hbk)
ISBN 978-3-0365-2559-4 (PDF)

Contents

About the Editor

Maria Carmela Di Piazza received her M.Sc. and Ph.D. degrees from the University of Palermo, Italy, both in Electrical Engineering. Since 2001, she has been a permanent staff research scientist with the National Research Council (CNR), Institute of Marine Engineering (INM). She has authored over 150 papers published in international journals and conference proceedings. She is currently a member of the editorial board of the journal Energies (Section A5: Smart Grids and Microgrids). She has led, for the CNR, several research projects and project activities, such as "Advanced Technologies for Energy Efficiency and Sustainable Mobility-OR2", "Innovative Electric Generation", and "TecBIA: Low Environmental Impact Technologies for the production of energy on ships-OR9". Her research interests cover shipboard electrical systems, renewable generation, power electronics and electrical drives, EMI/EMC, energy management systems (EMSs), and smart micro/nanogrids.

Preface to "Energy Management Systems for Optimal Operation of Electrical Micro/Nanogrids"

Energy management systems (EMSs) have been introduced in electrical power systems to perform optimized operations of the electrical grid infrastructure and to provide support to the grid operator in terms of optimized decisions. In electrical micro/nanogrids, the development of EMSs is crucial to correctly handle the uncertainties and intermittency of renewables. In addition, through their key functions (monitoring, control, optimization of flows, and use of electrical power), EMSs allow customers to play an active role in the energy market. The EMSs proposed so far were not conceived with respect to fostering their widespread and rapid adoption. For this to occur, several issues remain to be tackled: EMSs should be seamlessly integrated with the ecosystem of micro/nano grid components, including renewable generators, electrical storage systems (ESSs) and electrical loads; EMSs should interfere as little as possible with the comfort and habits of electricity market customers. In addition, the energy management algorithms should simultaneously provide advantages for both the end-user and the grid operator. On account of the previous considerations, this Special Issue collected papers addressing the development of EMSs specifically intended for the optimal operation of electrical micro/nanogrids and covering the following topics: optimization of electrical power flows in micro/nanogrids, EMSs for smart buildings, EMSs using vehicle integration in the microgrid, ESS modeling issues, forecasting for EMSs, and optimal use of ESSs in a microgrid.

Maria Carmela Di Piazza
Editor

energies

MDPI

Article

Performance Assessment of an Energy Management System for a Home Microgrid with PV Generation

Mahmoud Elkazaz [1,2], Mark Sumner [1,*], Seksak Pholboon [1], Richard Davies [1] and David Thomas [1]

1 Power Electronics, Machines and Control Research Group, The University of Nottingham, Nottingham NG7 2RD, UK; mahmoud.elkazaz@nottingham.ac.uk or mahmoud.elkazaz@f-eng.tanta.edu.eg (M.E.); seksak.pholboon@ttpi.tech (S.P.); ezzrrd@exmail.nottingham.ac.uk (R.D.); dave.thomas@nottingham.ac.uk (D.T.)

2 Department of Electrical Power & Machines Engineering, Tanta University, Tanta 31511, Egypt

* Correspondence: mark.sumner@nottingham.ac.uk

Received: 1 June 2020; Accepted: 1 July 2020; Published: 3 July 2020

Abstract: Home energy management systems (HEMS) are a key technology for managing future electricity distribution systems as they can shift household electricity usage away from peak consumption times and can reduce the amount of local generation penetrating into the wider distribution system. In doing this they can also provide significant cost savings to domestic electricity users. This paper studies a HEMS which minimizes the daily energy costs, reduces energy lost to the utility, and improves photovoltaic (PV) self-consumption by controlling a home battery storage system (HBSS). The study assesses factors such as the overnight charging level, forecasting uncertainty, control sample time and tariff policy. Two management strategies have been used to control the HBSS; (1) a HEMS based on a real-time controller (RTC) and (2) a HEMS based on a model predictive controller (MPC). Several methods have been developed for home demand energy forecasting and PV generation forecasting and their impact on the HEMS is assessed. The influence of changing the battery's capacity and the PV system size on the energy costs and the lost energy are also evaluated. A significant reduction in energy costs and energy lost to the utility can be achieved by combining a suitable overnight charging level, an appropriate sample time, and an accurate forecasting tool. The HEMS has been implemented on an experimental house emulation system to demonstrate it can operate in real-time.

Keywords: distribution systems; smart home; battery energy storage; energy forecasting; model predictive control; real-time control

1. Introduction

Home battery storage systems (HBSS) and home energy management systems (HEMS) can be of significant benefit to future electricity distributions by moving household electricity usage away from peak consumption times [1] and reducing the amount of local generation penetrating into the wider distribution system. This can also potentially help to defer the cost of grid re-enforcement associated with the increasing penetration of electric vehicles (EV), the electrification of heating, and the rapidly increasing use of domestic solar panels [2]. This can also lead to reduced electricity costs for the domestic consumer. For example, employing HBSS to capture surplus photovoltaic (PV) energy or off-peak utility energy to meet demand at peak-tariff times has been demonstrated in [3], and the use of demand side management (DSM) and the evolution of real-time pricing schemes also add to the capabilities of the HEMS to economically manage domestic electricity consumption [4,5].

Home energy management can be "optimized" using approaches such as model predictive [6], mixed-integer linear programming (MILP) [7], geometric programming, and dynamic programming [8].

For example, the authors in [9] used MILP optimization to manage a home with a HBSS, a PV array, and an EV with a "vehicle to home" option. A DSM strategy based on dynamic pricing and controlling power peaks was proposed in [10] which used a MILP-based model of the structure with an EV and an energy storage system.

Reference [11] presented a MILP-based HEMS together with an artificial neural network which forecasted residential loads. The energy management systems (EMS) and the forecasting model (using an artificial neural network (ANN)) employed the sample time of one hour for the load forecast; this is a very crude indication of the load profile as these profiles vary at a much faster rate. A rule-based EMS which aimed for optimized operation of a battery for use in electricity distribution grids with renewable energy sources (RES) has been proposed in [12]. The EMS maximized the use of the RES and prevented reverse power flow into the distribution transformer. Reference [12] controlled the battery considering only the current operating conditions without taking into account any potential changes in operating conditions—this could lead to impaired system performance.

To achieve an effective control for a HBSS based on predictions of load consumption and PV generation, [13] divided a household storage controller into two levels: a global control level and a local control level. The global algorithm is formulated and solved by convex optimization to determine future charging/discharging schemes for the storage system. Reference [14] proposed an alternative energy management scheme, integrating RES, electrical battery storage, and vehicle to grid. "Accurate" results are claimed, but clearly only running the algorithm once each day and using a sample time of one hour for management will lead to lower system performance due to the uncertainty of the generation and load demand.

Forecasting methods for PV generation and electricity consumption have been examined as part of several different studies. For example, in [15] a comprehensive analysis of PV prediction methods was presented which divided forecasting into deterministic and probabilistic methods. Most of those studies used data from historical measurements and/or weather forecasts. A recent literature review categorized demand forecasting models as statistic based or artificial intelligence-based models [16]. In [17], a forecasting algorithm for home demand was presented. The forecasting algorithm used a short sample time to forecast home consumption for one day ahead. To the best of our knowledge, only a few of these studies quantify the influence of these forecasting methods on the effectiveness of HEMS for PV-battery systems [18,19].

There is a gap in knowledge for designing HEMS derived using the analysis of real load and generation data obtained from electricity prosumers. The current literature is found to include many studies which examine PV-battery systems using poorly justified assumptions concerning the HBSS model (idealistic models which can lead to significant errors in the calculated system financial returns [20]) and/or datasets with a low sample resolution [11] (which result in errors in the system design and sizing, as sharp and rapid power changes are not taken into account).

Many of the HEMS introduced in the literature (e.g., [21,22]) have not considered the effects of forecasting uncertainties or different sample times on the economic performance of the HBSS or have ignored the effect of accurately adjusting the battery's overnight charging level [23]. Furthermore, the effect of a combination of different forecasting methods on PV-battery systems is not well understood. A review [24] suggested that the impact of forecasting on economic performance has not been studied in depth. Many studies quantify the operation of PV-battery systems by employing only one forecast method or assume a perfect forecast. The literature concludes that further investigation is required into the influence of forecasting for electricity demand and PV generation on the performance of PV-battery systems.

In addition, the selection of an appropriate overnight charging level for PV-battery systems has not been properly considered in the literature [25]. A limited number of studies considered overnight charging [26,27], but the battery was fully charged overnight (during the off-peak electricity tariff period) as they did not include any intelligent overnight charging control algorithms. Selecting an appropriate overnight charging level enhances the economic performance of PV-battery systems.

This paper presents a detailed investigation of a HEMS which employs both a real-time controller (RTC) and a model predictive controller (MPC). Their performance is evaluated in the presence of forecasting errors for different control sample times and for different HBSS overnight charging levels and different tariff policies. The HEMS presented here aims to minimize home energy costs, reduce energy lost to the supply utility, improve the local consumption of PV generation (self-consumption), and decrease the system dependency on external systems for forecasting. Two types of management strategies have been used: (a) energy management based on a RTC, and (b) energy management based on an MPC. A case study for a home in the UK is presented, which has typical household appliances, rooftop photovoltaic (PV) generation and a HBSS. The key contributions of this work are:

- This paper attempts to fill the gap in the literature by employing data for energy consumption and generation collected from *real* prosumers across the UK.
- It studies the importance of designing an HEMS which is able to respond quickly to changes in the system by operating with a short sample time (in this case two minutes), and analyses the resulting impact on the annual energy costs and the ratio of annual lost PV generated energy to the utility.
- It studies the performance of a HEMS which takes its own decisions locally while minimizing its dependence on external forecasting technologies (and complex communication infrastructures).
- It summarizes the requirements and challenges for HEMS and their impact on household energy costs; this can be considered an aid to selecting an appropriate controller for each PV-battery system.
- It studies the effect of forecasting errors, sample time resolution, tariff policies, the battery capacity and/or PV system on the performance of the MPC.

Experimental results for using an MPC-based HEMS are then presented to assess the performance of a real system.

The paper is organized as follows: Section 2 introduces the operating algorithm for the RTC-based HEMS and the influence of the charging level for the low tariff period overnight. Section 3 describes the operating algorithm of the MPC-based HEMS. This includes system modeling and the formulation of the optimization cost function (which is solved using a MILP approach). Section 4 introduces the specific cases analyzed in this paper. Section 5 shows the performance indicators which are used to assess the results obtained. Section 6 shows the simulation results obtained using RTC-based HEMS. Section 7 presents the experimental results obtained for MPC-based HEMS. Section 8 shows the annual performance analysis for MPC-based HEMS, and finally, Section 9 presents conclusions from this work.

2. Real-Time Controller-Based Energy Management System

For this system, the HEMS aims to minimize the daily household energy costs, reduce energy lost to the utility, and improve local PV self-consumption by controlling a HBSS using RTC. The RTC uses a rule-based algorithm to control the HBSS. During the off-peak tariff period (i.e., overnight period) the controller charges the HBSS to a preset overnight maximum charging level. During the rest of the day, the RTC discharges the HBSS—it compares the power at the point of grid connection and tries to make this power equal to zero. The main rules used for this controller are summarized as follows:

2.1. HBSS Discharging Mode

- If the household is drawing power from the supply utility at the point of grid connection, the HBSS tries to minimize the energy purchased by discharging the HBSS.
- If the household is drawing power from the supply utility and the power drawn is greater than the maximum discharge power of the HBSS, the HBSS will discharge at its maximum power and the remaining power will be purchased from the utility.
- If the household is drawing power from the supply utility and the HBSS state of charge (SOC) reaches its minimum value, the HBSS will cease discharging.

2.2. HBSS Charging Mode

The HBSS is charged (a) overnight when the purchase energy tariff from the utility is low, and (b) if the home is feeding power to the utility at the grid-connection point because excess PV energy is available.

2.2.1. Adjustment of the Low Tariff (Overnight) Charging Level

The HBSS is charged at night when the utility tariff is low. The overnight charging level is the maximum SOC that the battery should achieve during this period and should be adjusted according to operating conditions. For example, if the nighttime charging level has been set to a high value and the day ahead is sunny, the battery will be full and unable to receive any surplus PV energy during the day, which must therefore be exported to the utility (for little or no reward). On the other hand, if the next day is cloudy and the battery is not sufficiently charged during previous night, the battery may be completely discharged earlier than required and the household may have to buy energy from the supply utility at peak tariff prices. Five methods for adjusting the overnight charging level of the HBSS have been examined:

- **Constant Full Overnight Charging:** The battery charges fully during the off-peak tariff (i.e., night period from 12:00 to 7:00). There is no requirement to access the previous power profiles for load demand or PV generation. No weather forecasts or calendar timers are required [26,27].
- **Yearly Optimized Overnight Charging:** The battery is charged overnight to an optimized pre-set level (fixed throughout the year) depending on the battery capacity and the PV system size. This approach should yield better results than Constant Full Overnight Charging, as the battery is not always fully charged overnight and can be charged by any surplus PV generation.

To determine the optimal overnight charging level for the yearly optimized case, the operation of the system has to be simulated using historical data and different values for the overnight charging levels for one year to find the minimum annual household energy costs and the maximum annual PV self-consumption ratio. As can be seen from Appendix A, the point which achieves minimum annual household energy costs and maximum annual PV self-consumption ratio is the point at 80% overnight charging level. This point is selected to be the yearly optimized overnight charging level for the house under study. The same procedure is followed to determine the optimal overnight charging level for the season optimized case (described next).

- **Seasonal Optimized Overnight Charging:** Each season, the overnight charging level is adjusted to a different value. This value is selected based on the season and the PV and battery sizes. It is assumed that the HBSS contains a calendar timer to adjust the charging level at the beginning of each season. For example, for summer, the lowest charging level will be selected so that the HBSS is able to capture all excess PV generation during the next day.
- **Previous Day Modification:** The overnight charging level is adjusted based on the charging pattern for the previous day. For example, the overnight charging level increases by 10% for the current day if peak tariff energy was purchased during the previous day. The overnight charge level for the current day decreases by 10% if surplus PV energy was exported to the power grid the day before.
- **Weather prediction for the next day (i.e., next day PV generation forecasting):** Weather forecast data for the next day is used to adjust the overnight charging level of the battery which leaves capacity for the battery to be charged by the expected surplus PV generation the next day. Internet access is needed to download the meteorological forecast data for the next day and a PV forecasting model is needed to forecast the PV generation pattern. The weather forecast data is used to generate a forecasted PV generation pattern for the next day and then (1) is used to adjust the overnight charging level.

$$\text{Overnight charging level} = 1 - \frac{(1 - C_{PV}) \times E_{PVgen}{}^{expect}}{B_{Capacity}} \quad (1)$$

where $B_{Capacity}$ is the capacity of the battery (kWh), $E_{PVgen}{}^{expect}$ is the expected PV energy for the next day (this value is obtained using the forecasted PV generation pattern for the next day), and C_{PV} is the annual PV self-consumption ratio without the HBSS: this is the average value of the ratio of the total daily PV energy directly consumed in the home to the total daily generated PV energy; this value is obtained by simulating the system for one year without using the HBSS. This value is assumed to be fixed for the whole year.

This mode eliminates export and minimizes the amount of peak tariff energy purchased since the battery is topped up using any excess PV and off-peak energy. The authors in [15] listed several effective forecasting methods for PV generation for the day ahead.

2.2.2. Charging Using Excess PV Energy

If the home is feeding power to the utility when there is surplus PV energy, the following rules are used to charge the HBSS.

- If the home is feeding power to the utility, the HBSS charges to store the surplus energy.
- If the home is feeding power to the utility and this is greater than the HBSS maximum charging power, the HBSS will charge at its maximum charging power and the remaining power will be fed into the utility.
- If the home is feeding power to the utility and HBSS SOC reaches its maximum value, the HBSS will stop charging.

3. Model Predictive Control-Based Energy Management System

The MPC aims to optimize the control actions for the current sample. At each time step (t), the MPC performs an optimization process and computes an optimal control sequence for a finite horizon [28]. Only the first control action in the sequence is applied. Over the next time step (t + 1), the MPC receives new system measurements and recalculates the optimal control sequence for the next period.

In this paper, MILP optimization-based MPC is used to minimize the household energy costs, improve the self-consumption of PV generation and reduce energy lost through the control of the HBSS. The HBSS power settings obtained will ensure the best use of electrical energy. For every sample time, (1) forecasts for the profiles for PV generation and load demand over the next 24 h are obtained, (2) real-time measurements of the HBSS SOC are used to update the MPC, (3) MILP optimization is performed, and (4) the power references for the HBSS are updated. The time frame in which the MILP optimization is performed is t = 0:24 h. The optimization process is repeated every sample time (2 min). The HBSS control is optimized for subsequent time slots (from t = t + 1:24 h), noting that only the setting for the next time slot (t + 1) is sent to the HBSS.

3.1. *Formulation of the Optimization Problem and Constraints*

MILP optimization is used to minimize the household energy costs [29]. MILP is an approach to optimization which solves constrained optimization problems which include an objective function and a set of variables and constraints [30]. The formulation of the problem is defined as:

$$\text{Objective}: \ \text{minimize} = Cx$$

$$\text{Constraints}: \ A.x \leq b$$

$$\text{xmin} \leq x \leq \text{xmax}$$

where $x \in Z^n$ C, b are vectors and A is a matrix.

The objective function which needs to be minimized is the cost function in (2), which aims to minimize cost of energy and maximize the local use of the PV generation. The optimization finds the best solution to the objective function (2) from a set of potential solutions that meet the constraints, i.e., the equality constraints (5) and inequality constraints (9)–(15). A feasible solution is one that satisfies all constraints. The variables determined from the solution to the optimization problem are a set of optimal control settings "$P_{HBSS}(t)$" for the next 24 h with a two-minute resolution. These settings are then forwarded to the HBSS.

The daily household energy costs "C_{Home}" (2) that need to be minimized are comprised of payments (3) (e.g., for electricity purchased from the supply utility), and incomes (4) (e.g., for the energy exported to the supply utility) [31]. The constraints are divided into: (a) the equality constraint function (5), and (b) the inequality constraint functions (9)–(11).

$$C_{Home} = C_{Home\_buy} + C_{Home\_sell} \tag{2}$$

$$C_{Home\_buy} = \begin{cases} \sum_{to}^{T} \Delta T \times TR_{buy}(t) \times P_{Utility}(t), P_{Utility}(t) \geq 0 \\ 0, P_{Utility}(t) < 0 \end{cases} \tag{3}$$

$$C_{Home\_sell} = \begin{cases} \sum_{to}^{T} \Delta T \times TR_{sell}(t) \times P_{Utility}(t), P_{Utility}(t) < 0 \\ 0, P_{Utility}(t) \geq 0 \end{cases} \tag{4}$$

where C_{Home} is the daily household energy costs (£); C_{Home_buy} is the cost of the energy purchased from the supply utility (£), C_{Home_sell} is the revenue of the energy exported to the utility (£), ΔT is the sample time (h); $TR_{buy}(t)$ is the purchase tariff for electricity at time interval t (£/kWh), $TR_{sell}(t)$ is the sale tariff for electricity at time interval t (£/kWh), $P_{Utility}(t)$ is the electrical power drawn from the utility by the household at time interval t (kW): a negative value represents exporting power, whereas a positive value represents importing power.

(5)–(9) represent the model and the constraints of the home microgrid:

(5) describes the balance for the total active power in the home.

$$P_{Utility}(t) + P_{HBSS}(t) = P_{home_{load}}(t) - P_{PV_{gen}}(t) \tag{5}$$

where $P_{home\_load}(t)$ is the home's electrical load at time interval t (kW), $P_{PV\_gen}(t)$ is the power generated by the home PV system at time interval t (kW), and $P_{HBSS}(t)$ is the HBSS (battery + converter) power charged/discharged at time interval t (kW): a negative value denotes that the HBSS charges; a positive value denotes that the HBSS discharges.

The model of the HBSS is represented by (6) and (7):

$$E(t) = \begin{cases} E(t-1) - \frac{\Delta T \times P_{bat}(t)}{\eta_d}, P_{bat}(t) \geq 0 \\ E(t-1) - \Delta T \times \eta_c \times P_{bat}, P_{bat}(t) < 0 \end{cases} \tag{6}$$

$$SOC(t) = \frac{E(t)}{B_{Capacity}} \tag{7}$$

where $P_{bat}(t)$ is the power charged/discharged by the battery at time interval t (kW); $E(t)$ and $E(t-1)$ are the energy stored in the HBSS at times t and t − 1, respectively (kWh); η_d, η_c are the efficiencies of the battery when discharging and charging, respectively (%). $B_{Capacity}$ is the energy capacity of the battery (kWh), whilst $SOC(t)$ is the state of charge of the battery at time t (%).

(8) represents the power converter model. The power converter receives its instruction from the HEMS and is used to control the HBSS.

$$P_{HBSS}(t) = \begin{cases} P_{bat}(t) \times \eta_{Conv} \cdot P_{bat}(t) > 0 \\ \frac{P_{bat}(t)}{\eta_{Conv}}, P_{bat}(t) \le 0 \end{cases} \tag{8}$$

where η_{Conv} is the efficiency of the power converter (%).

The HBSS power constraint (9) defines the highest power ($P_{HBSS\ max}$) that can be discharged/charged by the HBSS.

$$-P_{HBSS\ max} \le P_{HBSS}(t) \le P_{HBSS\ max} \tag{9}$$

The HBSS SOC constraint (10) specifies the minimum and maximum SOC level of the HBSS. This constraint is used following the recommendation of the Institute of Electrical and Electronics Engineers (IEEE) [32], where the SOC constraints prevent deep discharge or overcharging of the HBSS to maximize the HBSS lifetime. Deep discharging and overcharging of the HBSS substantially reduce the battery life [33].

$$SOC_{min} \le SOC(t) \le SOC_{max} \tag{10}$$

where SOC_{max} and SOC_{min} are the SOC limits (%) of the HBSS.

The battery power is classified as charging power and discharging power. The following constraints (11)–(15) are used to enforce the connection restrictions and make sure that the HBSS power is unidirectional during each sample time.

$$\sigma_{disch}(t) + \sigma_{charg}(t) \le 1 \tag{11}$$

$$\sigma_{disch}(t) = \begin{cases} 1, P_{HBSS}(t) > 0 \\ 0, P_{HBSS}(t) \le 0 \end{cases} \tag{12}$$

$$\sigma_{charg}(t) = \begin{cases} 1, P_{HBSS}(t) < 0 \\ 0, P_{HBSS}(t) \ge 0 \end{cases} \tag{13}$$

$$P_{HBSS}{}^{disch}(t) \le \sigma_{disch}(t) \cdot P_{HBSS\ max} \tag{14}$$

$$P_{HBSS}{}^{charg}(t) \ge \sigma_{charg}(t) \cdot (-P_{HBSS\ max}) \tag{15}$$

$\sigma_{disch}(t)$ and $\sigma_{charg}(t)$ are binary variables that ensure the HBSS power flows in one direction for a particular sample time; $P_{HBSS}{}^{disch}(t)$ and $P_{HBSS}{}^{charg}(t)$ are the HBSS discharge and charge power, respectively, at time interval t (kW).

3.2. Forecasting Methods

The operation of the MPC requires the use of forecasting for load demand and PV generation. In this research, the load profile and PV generation profile forecasted for the next 24 h are used in the optimization process to find the optimal reference values for the HBSS. The following methods have been used to forecast the demand profile for the household for the next day:

- the previous day's load profile (L-PD).
- the previous week, same day load profile (L-PWSD).
- the average load profile of the previous week (L-AV).
- one of the load demand forecasting techniques (L-FP) of [34], such as ANN, auto regression integrated moving average (ARIMA)+ANN, adaptive neuro-fuzzy inference system (ANFIS) which show better results for demand forecasting.

For PV forecasting for the next day, three forecasting methods have been used:

- the previous day's PV generation profile (PV-PD).

- the average PV generation profile of the previous week (PV-AV).
- the next day's weather prediction data + PV forecasting model to determine accurately the forecasted PV pattern for the next day (PV-FP) [15]. The next day's forecasted PV profile can be received and updated every sample time. This forecasting method needs continuous internet access. This service is available from the utility company or a retail agent for an extra cost.

4. Case Study

The analysis undertaken is based around a typical UK house. It comprises common household appliances, rooftop PV generation and a HBSS. The house is connected to the supply utility. The household load profiles used are real measurements made in a UK based house [35]. This data is sampled with a one-minute resolution for a whole year. The total annual energy consumption for the home is 4104 kWh: this value is close to 4200 kWh which is the UK average for a medium sized house [36]. Measured data is also used for PV generation, obtained from the PVOutput.org website [37] for a 3.8 kW rooftop PV located in Nottingham. The data is for a full year with a sample time of one minute. The PV generation profile was scaled down to be equivalent to the PV generation of a 1.4 kW peak system, which was considered appropriate for the home under study.

Three electricity purchase tariff schemes were considered, namely: (a) Economy 7 (E7), (b) time of use (TOU), and (c) real-time pricing (RTP). The householders also have to pay a standing charge (24 pence per day) to account for distribution infrastructure costs. When selling surplus energy to the main utility, a fixed export sale price of 3.79 pence/kWh is used. The E7 purchase tariff values are from RobinHood Energy, UK [38]. The TOU purchasing tariff values are from Green Energy, UK [39]. The real-time pricing tariff values are derived from a dataset based on the total UK electricity consumption, available from New Electricity Trading Arrangements (NETA) [40], and lists the price per MWh associated with half hour timeslots. The export tariff values are from the Office of Gas and Electricity Markets (OFGEM) [41]. Figure 1 shows the different tariff schemes used in this research.

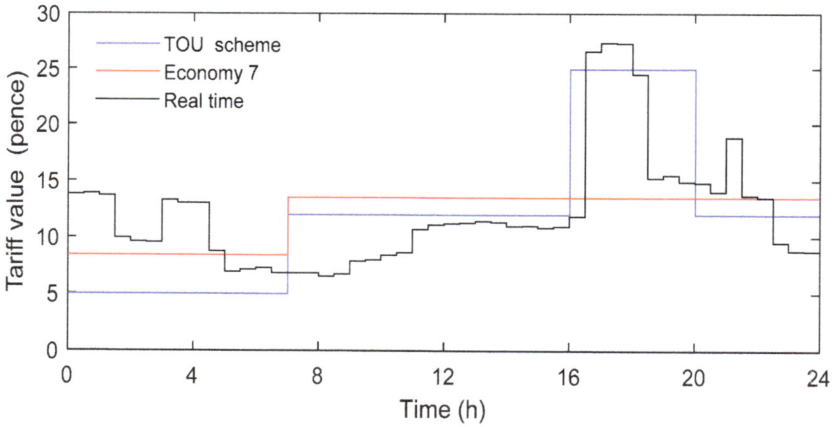

Figure 1. Values for Economy 7, time-of-use Tariff, and real-time pricing scheme.

The approach presented in [42] for determining the best size for an energy storage system was used to select an appropriately sized battery (in terms of energy and power rating) and to optimize the charging-discharging boundaries for the system presented in this paper. Investment costs were set at £135/kWh [43] for energy, £300/kW [41] for power. These investment costs include the installation cost of the HBSS. The parameters of the HBSS used in this research are shown in Table 1 [44,45].

Table 1. The parameters of the home battery storage systems (HBSS).

Battery Capacity	6.4 kWh
Battery efficiency (η_d, η_c)	95.3%
Rated power ($P_{HBSS\ max}$)	±2.5 kW
SOC_{min}	20%
SOC_{max}	90%
Converter efficiency (η_{Conv})	96.7%

5. Performance Indicators

Three performance indicators were used to quantify the performance of the HEMS:

- **Household energy cost increment ratio (HECIR):** The HECIR is the ratio of the actual household energy costs to the household energy costs that would be achieved in the ideal case, (16). If the value of the HECIR is 0, this means the system has ideal performance. As the value of the HECIR increases, this will indicate higher energy costs and lower system performance. The actual household energy costs are calculated using Equations (2)–(4). The ideal case for the household energy cost when using RTC will occur when the overnight charging level is determined accurately with zero forecasting error for PV generation. The ideal case for the household energy cost when using the MPC will occur when there is zero forecasting error for PV generation and load demand, and the lowest sample time of two minutes is used for the MPC implementation. For both of these cases the actual load/PV data is used for the forecast (i.e., there is an "ideal" forecast).

$$\text{HECIR} = \left(\frac{\text{Actual household energy cos ts}}{\text{Household energy cos ts (ideal case)}} - 1\right) \times 100 \tag{16}$$

- **PV self-consumption ratio (PVSCR):** This metric is used to calculate the quantity of the PV energy used in the home either directly or via the HBSS. The PVSCR is calculated by dividing the PV energy used in the house by the total PV energy generated, (17). A value of 100% indicates all the PV energy generated is used in the house and there is no export to the supply utility.

$$\text{PV self consumption ratio} = 1 - \frac{E_{PVgen}{}^{export}}{E_{PVgen}{}^{total}} \times 100 \tag{17}$$

where $E_{PVgen}{}^{total}$ is the total daily generated PV energy and $E_{PVgen}{}^{total}$ is the total daily exported PV energy to the main electricity grid.

- **Energy lost ratio (ELR):** The ELR is determined by dividing all the "lost energy" by the all PV energy generated (18). The "lost energy" is the exported energy to the supply utility because of (a) errors in forecasting, (b) larger sample times which lead to inaccurate power settings for the HBSS, (c) periods when the HBSS is fully charged and no further surplus energy can be stored. Ideally this lost energy should be stored in the battery to be used at peak tariff periods. This ratio is used to assess the performance of the MPC operation, with 0% meaning no lost energy. As the value of the ELR increases, more lost energy will accrue, leading to higher energy charges. The ELR index incorporates both the (unwanted) export resulting from inaccurate HBSS reference settings and from any surplus energy from the PV generation system: note that the complement of the PVSRC only quantifies the exported energy from the PV generation system during the day.

$$\text{Energy lost ratio} = \frac{E^{\ export}}{E_{PVgen}{}^{total}} \times 100 \tag{18}$$

where $E_{PV_{gen}}{}^{total}$ is the total daily PV energy generated and $E^{\ export}$ is the total daily energy exported to the main electricity grid.

6. RTC-Based HEMS—Results

6.1. Simulation Results for the RTC-Based HEMS for Two Days

The operation of the RTC-based HEMS was simulated over two days to help with understanding the real-time dynamic performance of the RTC-based HEMS. The simulation process used the rule-based control algorithm defined in Section 2, as well as the different adjustment techniques for the overnight charging level, to assess the daily performance of the RTC.

Figure 2 shows the performance of the RTC for two consecutive days using the following overnight settings. Case 1: constant full overnight charging, Case 2: yearly optimized overnight charging, Case 3: seasonal optimized overnight charging, Case 4: previous day modification, and Case 5: weather prediction for the next day. A new, Case 6 (Ideal case), was also created to be used as a reference case. Case 6 is similar to Case 5, the only change is that the PV generation forecast in Case 6 is assumed ideal, i.e., zero forecasting error.

In Case 1, it is clear from Figure 2(b)-case 1 that in each of the two days, the HBSS was charged up to its maximum limit (90%) during the night while the days were sunny, so that much of the surplus PV energy was exported to the grid, as shown in Figure 2(c)-case 1, and not stored in the HBSS as shown in Figure 2(a)-case 1. The HECIR and the PVSCR for the two days were 43.67% and 62%, respectively, which were poor.

In Case 2, a yearly optimized overnight SOC level was selected (i.e., 80%). Figure 2(b)-case 2 shows that the HBSS was charged up to 80% overnight and then supplemented with the surplus PV generation available during the day. The HECIR and the PVSCR for the two days were 35.6% and 70.8%, respectively, which was an improvement on case 1.

In Case 3, selecting a seasonal overnight charging setting allowed the HBSS to be charged by surplus PV energy through the day. This achieved a lower HECIR (14.7%) and higher PVSCR (92.8%) compared to case 1 and case 2. It is clear from Figure 2(c)-case 3, compared to (c)-case 1 and (c)-case 2, that the exported energy to the main electricity grid decreased, which means higher PVSCR. Generally speaking, for summer the best overnight charging level should be the lowest one to maximize the PVSCR. These settings ensure lower household energy costs and higher PVSCR, if appropriately sized HBSS and PV systems have been selected in advance. For smaller battery capacities, the best charging level over the four seasons was found to be the maximum available.

In Case 4, it is assumed that the overnight charging level set for the first day was 60% as can be observed from Figure 2(b)-case 4. The first day was sunny and surplus PV energy was exported to the grid as is clear from Figure 2(c)-case 4. The RTC decreased the overnight charging level for the second day to 50% (i.e., decrease by 10%) to reduce the exported PV energy during the second day. The HECIR and the PVSCR for these two days were found to be 15.88% and 92.8%, respectively—these values are similar to the values observed in case 3.

In Case 5, weather prediction for the next day was used to accurately adjust the overnight charging level. For the house under study, it can be seen that 65% of the total generated PV energy was directly used in household consumption without contribution from the HBSS. The overnight charging level for each day was adjusted according to (1). The overnight charging levels for the two days were 48% and 52.5%, respectively, as can be seen in Figure 2(b)-case 5. The HECIR and the PVSCR for these two days were found to be 4.41% and 96.7%, respectively. It is clear from Figure 2(c)-case 5 that accurately adjusting the overnight charging level for each day minimizes the exported excess PV energy and maximizes the PVSCR.

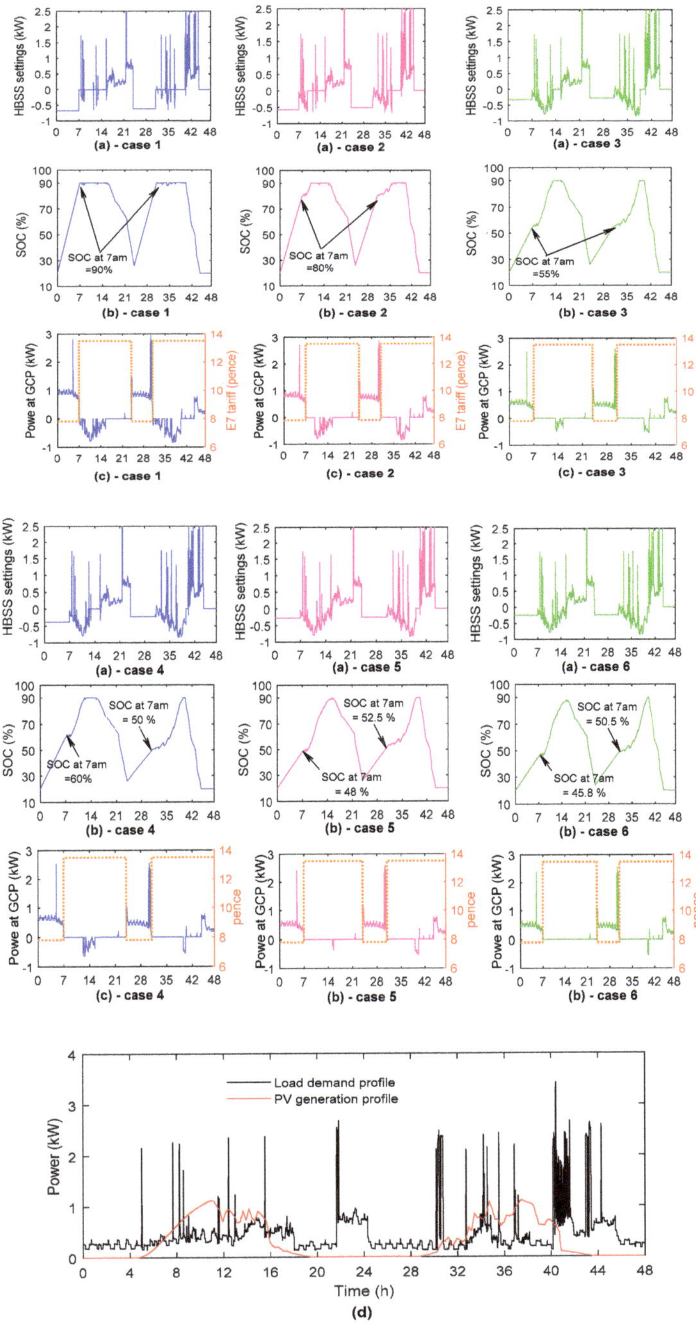

Figure 2. The performance of the real-time controller (RTC)-based home energy management systems (HEMS) for two consecutive days using Cases 1–6, respectively; (**a**) the HBSS power settings obtained from the RTC, (positive—HBSS is discharging, negative—HBSS is charging); (**b**) the resultant state of charge (SOC) curve of the HBSS; (**c**) the resultant power from the supply utility, (positive—house is importing power from the utility, negative—exporting) and the associated E7 tariff values; (**d**) the household consumption and PV generation profiles for two consecutive days.

In Case 6, the ideal case, it is assumed that the PV generation for the next day was known perfectly (which is possible as we were using historic data) and was used to accurately adjust the overnight charging level (i.e., as discussed in case 5). This case is used as a reference case. The HECIR and the PVSCR, in this case, were found to be 0.68% and 98.8%, respectively—almost perfect.

6.2. Annual Performance Analysis for the RTC-Based HEMS

The performance of the RTC-based HEMS was then tested for a one-year period to consider the yearly financial effect and to consider all four seasons of the year. This section assesses how the annual household energy costs and the annual PVSCR were affected using the five overnight charging modes, i.e., discussed in Section 2.2.1. Table 2 shows the annual HECIR and the annual PVSCR using the different overnight charging levels. The simulation results obtained in this section are for a full year to take into consideration all the seasons of the year.

Table 2. The annual Household energy cost increment ratio (HECIR) and the annual PV self-consumption ratio (PVSCR) obtained using different overnight charging levels while using the E7 purchasing tariff.

Case	Overnight Charging Mode	Annual HECIR (%)	Annual PVSCR (%)
1	Constant full	25	42.7
2	Yearly optimized	19.4	59.5
3	Season optimized	15.3	66.6
4	Previous day modification	13.54	71.7
5	Weather prediction for the next day * (i.e., 14% MAPE)	8.1	89.70
6	Ideal case (the actual PV generation profile of the current day is used)	-	94.1

* The next day PV forecast using the weather prediction is discussed more in Section 8.2.

The results presented in Table 2 are for the case in which the RTC-based HEMS is used to manage the household energy. No forecasted load demand or PV generation profiles were required in any of the five cases in Table 2 as RTC-based HEMS depends on the real measurements and a rule-based algorithm rather than predicated profiles to determine the HBSS settings for each time step. Only in case 4 is the PV generation forecast for the next day used (using weather prediction for the next day) but only to adjust the overnight charging level of the battery.

In case 4, the forecasted PV generation would normally be obtained for one time only (it is not updated at each time step) using the meteorological forecast data for the next day and a PV forecasting model. In this work, as historical data is being used, the forecasted PV generation profile was created by adding Gaussian noise to the actual PV generation profile of the current day. The Gaussian noise represents the MAPE for the forecasted profile. The value of the MAPE (14% in this case) was obtained from the results available from the Sheffield solar website for the forecasting of PV generation for the next day [46].

It can be seen from the results in Table 2 that accurate adjustment of the overnight charging level for the HBSS is very important and affects both the annual home energy savings and the PV self-consumption. If the appropriate overnight charging level is selected for each season (i.e., as in case 3), a lower home energy cost is achieved compared to case 2 and case 1. Case 5 (i.e., weather prediction for the next day) achieves the lowest annual HECIR compared to the other cases. It is also worth noting that in case 5, a continuous connection to the internet is required to download the weather forecast for the next day to be able to determine the overnight charging level of the HBSS. Additional costs may be required for a contract for a suitable forecasting package that updates the system with up-to-date weather prediction data.

7. MPC Based HEMS—Experimental Results

A laboratory system has been constructed to evaluate the performance of the MPC using a real HBSS in a typical operating environment. The MPC-based HEMS was tested experimentally for one day at the FlexElec Laboratory in the University of Nottingham, using the "Smart Home Rig" (SHR) shown in Figures 3 and 4.

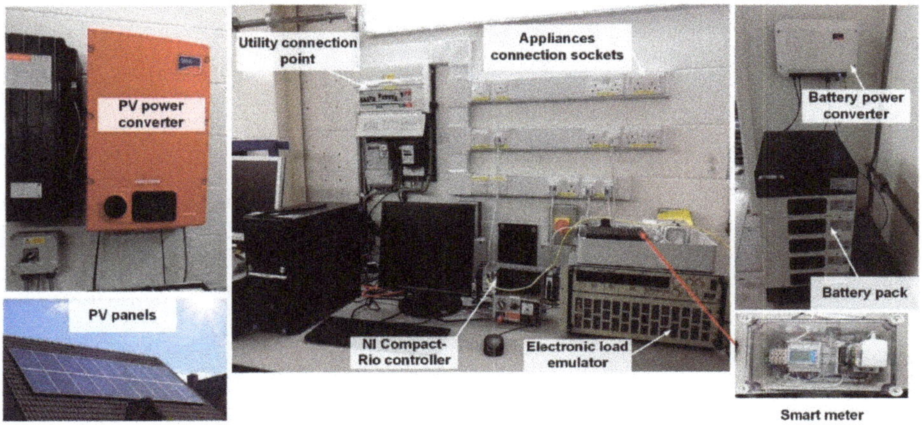

Figure 3. The smart home rig at the University of Nottingham Laboratory.

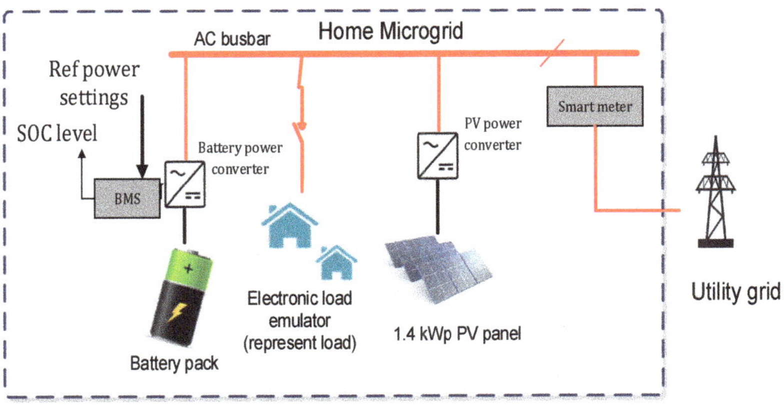

Figure 4. The connection diagram of the smart home rig at the University of Nottingham Laboratory.

This SHR comprises:

- Home battery storage system comprising (a) BYD lithium-ion battery pack, 6.4 kWh [44] and (b) SMA bidirectional power converter, 2.5 kW [45].
- 1.4 kWp PV system with a 3.68 kW SMA PV inverter [47]. The PV solar panels are located on the rooftop of the FlexElec laboratory.
- ZSAC Electronic AC load emulator, 5.6 kW [48]: the programmable load emulator receives the digital load demand profiles and creates a real current/power profile drawn from one of the appliance sockets in the SHR. LabVIEW software and a NI CRio FPGA system [49] extract the numerical load values from the database and send them to the programmable load emulator as a reference value.

- Smart meter: a three-phase smart meter used to measure PV generation, load demand, and the power imported/exported by the house from/to the supply utility. The smart meter uses a two minute sample time [50].
- PC: Core i3-7100 CPU, 3.91 GHz PC: the PC is used to run the HEMS.
- Raspberry Pi: used as a Modbus communication interface between the smart meter and the battery management software on the PC. It is also used as a communication interface between the battery management software on the PC and the battery power converter to send the optimal power settings to the SMA converter of the HBSS, and read the actual SOC of the battery.
- Software used: (a) MATLAB—to execute the optimization algorithm and perform the forecasting process, and (b) LABVIEW software package—to control the programmable load emulator.

The HEMS-based MPC was implemented experimentally. At each sample time (every two minutes): (1) the Raspberry pi measures the SOC of the HBSS (from SMA converter in the HBSS) and sends it to the HEMS; (2) a MATLAB script is used to execute the MILP optimization and calculate the optimal power setting for the HBSS; (3) the Raspberry Pi receives the HBSS optimal power setting for just the next sample and passes it to the HBSS's SMA inverter; (4) these steps are repeated every two minutes.

Figure 5 shows the performance of the MPC-based HEMS for one day. The TOU tariff scheme and a fixed export electricity tariff were used in this experiment. The methods used for forecasting demand and generation are the previous week same day load profile (L-PWSD) and the previous day generation profile (PV-PD), respectively. The mean absolute percentage error (MAPE) for the load and generation forecasts were 29.3% and 22.66%, respectively. A two-minute sample time has been used—the MPC updates the HBSS references every two minutes and it can therefore respond to relatively fast disturbances in the system.

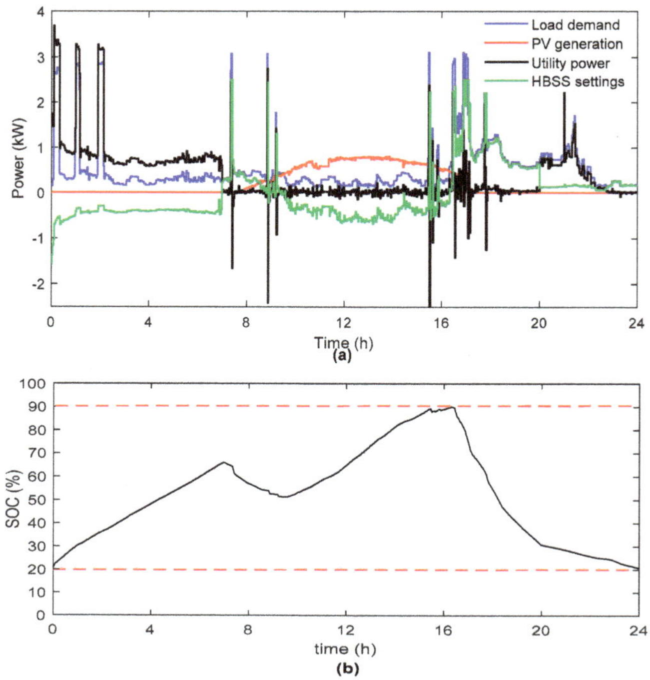

Figure 5. (**a**) Actual daily, load demand, PV generation, utility power (negative value—house is exporting power to the utility, positive value—importing), and the optimal power settings sent to the HBSS (negative value—charging, positive value—discharging); (**b**) Daily actual SOC of the HBSS.

A sample time of two minutes is the shortest sample time that can be used in this experiment. When a one-minute sample time was attempted for MPC operation, it was found that the MPC takes 5.62 min to perform just the optimization process, making a sample time of less than two minutes unfeasible for this experiment.

Figure 5a shows that the HEMS/BESS matches the household demand from 16:00 to 20:00 (during peak-tariff hours) so the home did not have to import energy from the main utility during this period. The PV generation was used in the home (including charging the HBSS) instead of being exported to the utility. From 00:00 am to 07:00 am (off-peak tariff time), a greater amount of energy was drawn from the supply utility at the low tariff rate (5 pence/kWh) to cover the home energy demands and charge the HBSS. It is clear from Figure 5b that the HBSS was charged from both the surplus PV generation during daytime and the imported energy from the supply utility during the off-peak tariff time.

Unwanted export power can be seen in Figure 5a (negative values of the utility power (black) profiles). The reason for this unwanted export was the errors associated with the load and generation forecasts at certain points in the day (i.e., when there is a sudden increase or decrease of the load or generation/export at power levels higher than the BESS can manage). The unwanted export power was one of the reasons for the lost energy when using the MPC for HEMS. The HECIR and ELR were 27% and 14%, respectively.

It is clear from Figure 5b that the HBSS charged to 67% overnight (i.e., not to its maximum limit of 90%) because this overnight charging level (a) enables the HBSS to provide the expected load demand during the morning period (i.e., no energy is purchased from the supply utility from 7:00 to 10:00), and (b) leaves space for the surplus PV generation during the following day to be stored in the HBSS (i.e., no energy is exported to the main utility from 9:00 to 15:00). The battery is fully charged at 16:00hrs, ready for the peak tariff period.

8. Performance Analysis for the MPC-Based HEMS

This section will analyse the performance of the MPC-based HEMS over a one-year period.

8.1. Sample Time Resolution

With a sample time of 60 min for the MPC operation, the HBSS power settings received from the MPC optimization will stay fixed for 60 min. As a result, any change in generation and/or load in this period will be compensated by the supply utility to balance the total active power in the home (4) and this may affect the total energy costs. If a sample time of two-minutes is selected for the MPC, it will update the HBSS references every two minutes and therefore will respond to fast changes in load and generation to minimize the home's energy costs and reduce lost energy.

Figure 6 compares the use of a 60 min sample time and a two minute sample time for the MPC. Figure 6a shows the HBSS power settings obtained using 60 min (red settings) and two minutes (blue settings), respectively. Figure 6b shows the power drawn from the supply utility when using 60 min (red) and two minutes (blue), respectively. Figure 6c shows the load and the generation profiles for a two-minute sample time.

It can be seen from Figure 6a that when a 60 min sample time was used for the MPC, the HBSS references remained constant for 60 min and changes in load and generation were compensated by the supply utility, as can be seen in Figure 6b. Energy is purchased from the supply utility during the peak-tariff period, and there is also unwanted export to the supply utility during the late afternoon. This export could be captured in the HBSS. Figure 6a shows that when the MPC updates the HBSS power settings (blue line) every two minutes, it can respond appropriately to fast changes in load and generation (seen in Figure 6c).

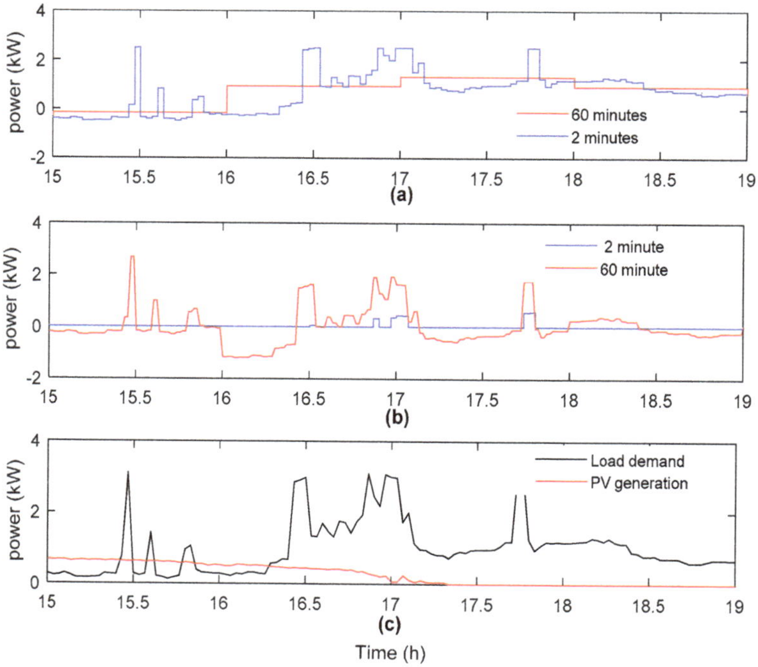

Figure 6. Comparison of a two-minute sample time and a 60 min sample time for the MPC. (**a**) The HBSS optimal power settings when using a two-minute sample time (blue settings) and when using a 60 min sample time(red), (**b**) the power drawn from the supply utility when using a two-minute sample time (blue) and a 60 min sample time (red), (**c**) PV generation and load demand profiles for a two-minute sampling time.

Table 3 shows the effect on the operation of the MPC using different sample times. An ideal forecast of load and generation was used for these tests so that the effect of sample time only was studied. The best case is where a two-minute sample time was used.

Table 3. Effect of sample time on the MPC computational time, the annual HECIR, and the annual energy lost ratio (ELR).

Sampling Time Resolution (min)	HECIR (%)	ELR (%)	MPC Computation Time (s)
60	35.19	29.86	4.99
30	26.31	24.1	5.81
15	21.7	19.95	6.23
5	10.69	10.86	11.3
2	0	5.9	95.1
1 *	-	-	337.5

* This case cannot be applied in a real system. The optimization process was only performed to show the required computation time.

It can be seen that the computational time of the optimization process can pose a problem if too short a sample time is used. For example, it is seen that if a one min sample time is used, the MPC takes 5.62 min to perform just the optimization process (a much larger time for computation than the rolling step size itself). This makes the use of this sampling time resolution unfeasible. In addition, if a very short sample time is used, this will force the controller to respond to each and every change in the load or generation. The controller action then has a high frequency content which can affect the lifetime of the HBSS: the battery will be exposed to high operational stresses if it changes between charging and discharging too quickly.

Usually, it is desirable to use a short sample time for the MPC. It can be seen from Table 3 that when a small sample time is used, this results in a lower energy cost increment ratio and a lower lost energy ratio. For a 60 min sample time for the MPC, the HECIR increases by 35.19% and the lost energy ratio increases by 29.86% compared with the smaller sample time. If a short scanning and response time is used, the MPC controller can respond to rapid changes in load and generation, and this therefore guarantees better performance and a greater reduction in costs for the householders. The compromise is that a longer computation time is required for the MPC optimization process.

8.2. The Effect of Forecasting Errors

To measure the accuracy of the forecasting methods for load and generation for the following day, the mean absolute percentage error (MAPE) is calculated (19).

$$\text{M.A.P.E} = \frac{1}{N}\sum_{to}^{T}\left|\frac{A_t - F_t}{A_t}\right| \times 100 \tag{19}$$

where A_t is the actual point, F_t is the forecast and N is the number of forecasts considered.

Table 4 shows the MAPE values for the forecasted load and generation for the next day using the forecasting methods listed in Section 3.2. The forecasted load and generation profiles using the methods (i.e., L-PD, L-PWSD, L-AV, PV-PD, and PV-AV) were found using the historical dataset available.

Table 4. The mean absolute percentage error (MAPE) values for the load demand and the PV generation forecasting methods listed in this research.

Forecasting Method	L-PD	L-PWSD	L-AV	L-FP	PV-PD	PV-AV	PV-FP
MAPE (%)	39.6	34.3	45.5	29.85	25.45	29.9	14

When using the PV-FP forecasting method, as historical data has been used, the forecasted PV generation profile was created by adding Gaussian noise to the actual PV generation profile of the current day. The Gaussian noise represents the MAPE for the forecasted profile. The value of the MAPE (14% in this case) is obtained from the results available from the Sheffield solar website for PV generation forecasting for the following day [46].

For the next day household demand forecasting using the L-FP case, an adaptive neuro-fuzzy inference system (ANFIS) forecasting method, developed in [51], was used. Other load demand forecasting techniques could be used to potentially obtain better results.

Figure 7 shows the effect of forecasting uncertainty for both the load and generation on the annual HECIR and the annual lost energy ratio using the TOU purchasing tariff scheme. The sample time used in these calculations is two minutes and is fixed in all the cases to investigate the effect of the forecasting uncertainty only. It can be seen from the results that the forecasting uncertainty for the load demand and PV generation for the following day greatly affect the household energy costs and the lost energy ratio. The HECIR approaches 67.98% when using the L-AV and PV-AV forecasting methods (i.e., more than half of the cost that would be achieved when using ideal forecasting). Ideal forecasting is the perfect forecasting (100% accurate) of generation and load profiles for the next 24-h period, which can be created as we are using historical data profiles. The ELR also approaches 32.33% for the same case. This lost energy should be saved in the HBSS and used at the appropriate time period rather than being lost to the utility with little reward. From Figure 7 it can be seen that using a forecasting method such as L-FP and PV-FP achieves lower HECIR and ELR. It is worth noting that the actual HECIR and ELR will be higher than the values shown in Figure 7 if a longer sample time is used for the MPC.

(a)

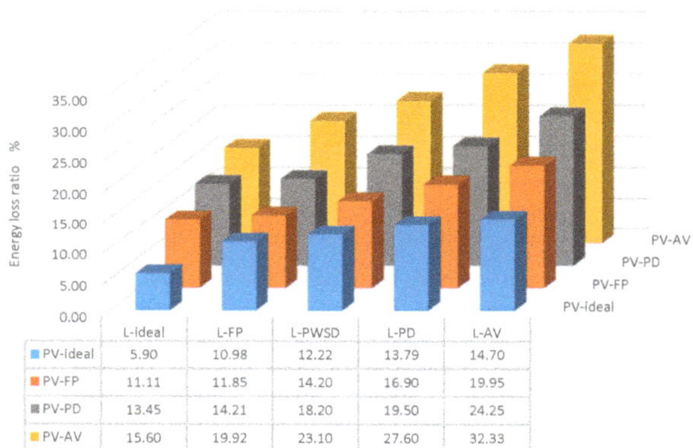

(b)

Figure 7. The effect of forecasting uncertainty for both the load demand and PV generation on (**a**) the annual HECIR and (**b**) the annual ELR, using the time of use (TOU) purchasing tariff scheme and two-minute sample time.

8.3. The Effect of Changing Tariff

Table 5 shows the annual household energy costs calculated using the three purchasing tariff schemes shown in Figure 1. The sample time used in this section is two minutes and perfect forecasting was used for both the load demand and PV generation as we are using historical data.

Table 5. The annual household energy costs for the three purchasing tariff schemes.

Purchasing Tariff Scheme	Annual Household Energy Costs (£)
Economy 7	347.2
Time of use	298
Real time pricing	327.4

It can be seen from the results in Table 5 that lower home energy costs can be achieved using the TOU tariff compared to using the Economy 7 or the real-time tariff schemes. The TOU tariff offers lower energy prices during off-peak periods (i.e., 4.99 pence/kWh as shown in Figure 1), compared to the Economy 7 tariff (8.4 pence/kWh for the same off-peak period). Lower energy prices during off-peak periods give the HBSS a chance to store as much energy as needed at low cost to cover the home demands through the day. The TOU tariff also offers lower prices during the off-peak periods compared to the real-time (half-hourly) pricing scheme which can offer high prices at night (as can be seen in Figure 1). This is an area of ongoing research.

8.4. Variation of HBSS Capacity

Table 6 shows the impact of changing the capacity of the HBSS on the annual household energy cost and the PV self-consumption ratio using the TOU tariff scheme and 1.4 kW peak PV system. It can be observed that as the battery capacity increases, the household energy costs decreases and the PVSCR increases. The PVSCR also increases at a high rate when the battery capacity changes from 0 kWh to 4.8 kWh. However, the increment rate in the PVSCR is low when the battery capacity increases from 6.4 to 13.5 kWh. This is related to the rated size of the PV system (1.4 kWp); when the battery capacity increases beyond a certain size, this additional storage capability cannot increase the capture of PV generation (the remaining excess PV generation is at a power level above the power rating of the HBSS), and it therefore cannot improve the PVSCR.

Table 6. Effect of changing battery storage's capacity on the annual household energy cost and the PV self-consumption ratio using TOU purchasing tariff scheme and 1.4 kW (peak) PV system.

Battery Capacity (kWh)	Annual Household Energy Costs (£)	PVSCR (%)
13.5	240.71	91.56
9.6	264.88	89.88
6.4	298	88.47
4.8	322.83	87.23
2.5	352.6	84
No battery	393.51	61

It is worth noting that as the battery capacity increases, the initial investment cost of the battery system increases as well. An optimization technique is required to select the best battery size which minimizes both battery investment cost and the annual household energy costs, as discussed in Section 4 and [42].

8.5. Varying PV System Size

Table 7 shows the effect of changing the size of the PV system on the annual household energy cost and PV self-consumption ratio using the TOU purchasing tariff scheme and a 6.4 kWh battery. Different PV system sizes, from 1 kW to 5 kW, were used by scaling the PV data accordingly. It is clear from Table 7 that as the PV system size increases, the household energy costs decreases. Furthermore, it is observed that the PVSCR decreases instead of moving to 100% as the PV system size increases. The reason for this is due to the battery power limit (2.5 kW in this case); the additional PV generation is at a power level higher than the battery system's converter and therefore much of the surplus PV energy is exported to the main electricity grid.

The appropriate PV system size for the house should be selected according to the household needs and in coordination with the power limits of the HBSS to improve the PVSCR and minimize the overall household energy costs.

Table 7. Effect of changing PV system size on the annual household energy cost and the PV self-consumption ratio using TOU purchasing tariff scheme and 6.4 kWh battery.

PV System Size (kW) Peak	Annual Household Energy Costs (£)	PVSCR (%)
5	74	42.6
3.5	160.1	55.63
2.5	221.3	68.35
1.4	298	87.47
1	330.4	92.9
No PV system	440.3	-

9. Conclusions

This paper has assessed the performance of two home energy management systems based on (a) a real-time controller and (b) a model predictive controller over a one-year period. Using the real-time controller, the effect of adjusting the overnight charging level on the overall performance has been studied. The results showed that the lowest value for household energy cost increment ratio and the highest value for PV self-consumption ratio (i.e., 8.1% and 89.70%, respectively) could be achieved using a weather prediction for the next day to adjust the overnight charging level, but this would incur additional operational costs.

Load demand and PV generation forecasts can be made relatively easily using methods such as L-PWSD, L-PD, L-AV, PV-PD, and PV-AV, i.e., methods which use historical data only and do not require any complex forecasting model or meteorological data (i.e., temperature, irradiation, humidity, etc.), compared to using accurate prediction methods such as L-FP and PV-FP which require up-to-date weather prediction and complex modelling. L-FP and PV-FP forecasting packages achieve greater reductions in household energy costs and lower lost energy compared to simple prediction packages. However, these forecasting packages require a good communication infrastructure and also additional costs for complex modelling.

The performance of the MPC has been studied considering the effect of forecasting errors (this technique requires forecasting for its fundamental operation), the sample time, and different purchasing tariffs. The results show that with appropriate selection of the forecasting method for load demand and PV generation, a significant reduction in household peak energy demand from the supply utility and also the cost of home utility bills can be achieved. Using a 60 min sample time for MPC operation increases the household energy cost increment ratio by 35.2% and the lost energy ratio by 29.8% compared to using a two-minute sample time. Using a short scanning and response time of two minutes, the MPC controller can respond to changes in load and generation that occur over a short time, and therefore guarantees better performance and a higher reduction in costs for the householders. Using the time of use tariff scheme with a PV-battery system reduces the household energy costs even further compared to the other tariff schemes considered.

Author Contributions: Writing—original draft preparation, investigation and analysis, M.E.; methodology, M.S.; conceptualization, M.S. and D.T.; simulation, M.E.; experimental validation, M.E.; writing—review and editing, M.S. and D.T.; Visualization and Investigation, S.P.; validation and Software, R.D. All authors have read and agreed to the published version of the manuscript.

Funding: This work is supported by the University of Nottingham, the Egyptian Government—ministry of higher education (cultural affairs and missions sector) and the British Council through Newton-Mosharafa fund.

Conflicts of Interest: The authors declare no conflict of interest.

Appendix A

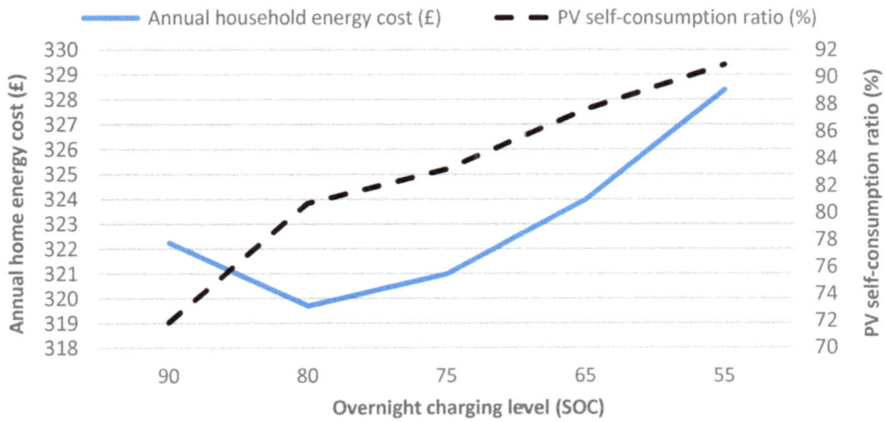

Figure A1. The annual household energy costs and the annual PV self-consumption ratio using different overnight charging levels for the yearly optimized case (case 2).

References

1. Chandra, L.; Chanana, S. Energy Management of Smart Homes with Energy Storage, Rooftop PV and Electric Vehicle. In Proceedings of the 2018 IEEE International Students' Conference on Electrical, Electronics and Computer Science (SCEECS), Bhopal, India, 24–25 February 2018; 2018; pp. 1–6.
2. Mehdi, L.; Ouallou, Y.; Mohamed, O.; Hayar, A. New Smart Home's Energy Management System Design and Implementation for Frugal Smart Cities. In Proceedings of the 2018 International Conference on Selected Topics in Mobile and Wireless Networking (MoWNeT), Tangier, Morocco, 20–22 June 2018; 2018; pp. 149–153.
3. Lebrón, C.; Andrade, F.; O'Neill, E.; Irizarry, A. An intelligent Battery management system for home Microgrids. In Proceedings of the 2016 IEEE Power & Energy Society Innovative Smart Grid Technologies Conference (ISGT), Minneapolis, MN, USA, 6–9 September 2016; 2016; pp. 1–5.
4. Terlouw, T.; AlSkaif, T.; Bauer, C.; van Sark, W. Multi-objective optimization of energy arbitrage in community energy storage systems using different battery technologies. *Appl. Energy* **2019**, *239*, 356–372. [CrossRef]
5. Van Der Stelt, S.; AlSkaif, T.; van Sark, W. Techno-economic analysis of household and community energy storage for residential prosumers with smart appliances. *Appl. Energy* **2018**, *209*, 266–276. [CrossRef]
6. Sun, C.; Sun, F.; Moura, S.J. Nonlinear predictive energy management of residential buildings with photovoltaics & batteries. *J. Power Sources* **2016**, *325*, 723–731.
7. Wang, G.; Zhang, Q.; Li, H.; McLellan, B.C.; Chen, S.; Li, Y.; Tian, Y. Study on the promotion impact of demand response on distributed PV penetration by using non-cooperative game theoretical analysis. *Appl. Energy* **2017**, *185*, 1869–1878. [CrossRef]
8. Muratori, M.; Rizzoni, G. Residential demand response: Dynamic energy management and time-varying electricity pricing. *IEEE Trans. Power Syst.* **2015**, *31*, 1108–1117. [CrossRef]
9. Erdinc, O. Economic impacts of small-scale own generating and storage units, and electric vehicles under different demand response strategies for smart households. *Appl. Energy* **2014**, *126*, 142–150. [CrossRef]
10. Erdinc, O.; Paterakis, N.G.; Mendes, T.D.; Bakirtzis, A.G.; Catalão, J.P. Smart household operation considering bi-directional EV and ESS utilization by real-time pricing-based DR. *IEEE Trans. Smart Grid* **2014**, *6*, 1281–1291. [CrossRef]
11. Paterakis, N.G.; Taşcıkaraoğlu, A.; Erdinc, O.; Bakirtzis, A.G.; Catalao, J.P. Assessment of demand-response-driven load pattern elasticity using a combined approach for smart households. *IEEE Trans. Ind. Inform.* **2016**, *12*, 1529–1539. [CrossRef]
12. Purvins, A.; Sumner, M. Optimal management of stationary lithium-ion battery system in electricity distribution grids. *J. Power Sources* **2013**, *242*, 742–755. [CrossRef]

13. Wang, Y.; Lin, X.; Pedram, M. Adaptive control for energy storage systems in households with photovoltaic modules. *IEEE Trans. Smart Grid* **2014**, *5*, 992–1001. [CrossRef]
14. Melhem, F.Y.; Grunder, O.; Hammoudan, Z.; Moubayed, N. Optimization and energy management in smart home considering photovoltaic, wind, and battery storage system with integration of electric vehicles. *Can. J. Electr. Comput. Eng.* **2017**, *40*, 128–138.
15. Antonanzas, J.; Osorio, N.; Escobar, R.; Urraca, R.; Martinez-de-Pison, F.J.; Antonanzas-Torres, F. Review of photovoltaic power forecasting. *Sol. Energy* **2016**, *136*, 78–111. [CrossRef]
16. Khan, A.R.; Mahmood, A.; Safdar, A.; Khan, Z.A.; Khan, N.A. Load forecasting, dynamic pricing and DSM in smart grid: A review. *Renew. Sustain. Energy Rev.* **2016**, *54*, 1311–1322. [CrossRef]
17. Gajowniczek, K.; Ząbkowski, T. Electricity forecasting on the individual household level enhanced based on activity patterns. *PLoS ONE* **2017**, *12*, e0174098. [CrossRef] [PubMed]
18. Moshövel, J.; Kairies, K.-P.; Magnor, D.; Leuthold, M.; Bost, M.; Gährs, S.; Szczechowicz, E.; Cramer, M.; Sauer, D.U. Analysis of the maximal possible grid relief from PV-peak-power impacts by using storage systems for increased self-consumption. *Appl. Energy* **2015**, *137*, 567–575. [CrossRef]
19. Riesen, Y.; Ballif, C.; Wyrsch, N. Control algorithm for a residential photovoltaic system with storage. *Appl. Energy* **2017**, *202*, 78–87. [CrossRef]
20. Shakeri, M.; Shayestegan, M.; Reza, S.S.; Yahya, I.; Bais, B.; Akhtaruzzaman, M.; Sopian, K.; Amin, N. Implementation of a novel home energy management system (HEMS) architecture with solar photovoltaic system as supplementary source. *Renew. Energy* **2018**, *125*, 108–120. [CrossRef]
21. Beaudin, M.; Zareipour, H. Home energy management systems: A review of modelling and complexity. *Renew. Sustain. Energy Rev.* **2015**, *45*, 318–335. [CrossRef]
22. Mohammadi, S.; Momtazpour, M.; Sanaei, E. Optimization-Based Home Energy Management in the Presence of Solar Energy and Storage. In Proceedings of the IEEE 21st Iranian Conference on Electrical Engineering (ICEE), Mashhad, Iran, 14–16 May 2013; pp. 1–6.
23. Gitizadeh, M.; Fakharzadegan, H. Battery capacity determination with respect to optimized energy dispatch schedule in grid-connected photovoltaic (PV) systems. *Energy* **2014**, *65*, 665–674. [CrossRef]
24. Suganthi, L.; Samuel, A.A. Energy models for demand forecasting—A review. *Renew. Sustain. Energy Rev.* **2012**, *16*, 1223–1240. [CrossRef]
25. Ibrahim, I.A.; Khatib, T.; Mohamed, A. Optimal sizing of a standalone photovoltaic system for remote housing electrification using numerical algorithm and improved system models. *Energy* **2017**, *126*, 392–403. [CrossRef]
26. Chen, S.; Gooi, H.B.; Wang, M. Sizing of energy storage for microgrids. *IEEE Trans. Smart Grid* **2011**, *3*, 142–151. [CrossRef]
27. Barnes, A.K.; Balda, J.C.; Geurin, S.O.; Escobar-Mejía, A. Optimal Battery Chemistry, Capacity Selection, Charge/Discharge Schedule, and Lifetime of Energy Storage Under Time-of-Use Pricing. In Proceedings of the 2nd IEEE PES International Conference and Exhibition on Innovative Smart Grid Technologies, Manchester, UK, 5–7 December 2011; pp. 1–7.
28. Shaikh, P.H.; Nor, N.B.M.; Nallagownden, P.; Elamvazuthi, I.; Ibrahim, T. A review on optimized control systems for building energy and comfort management of smart sustainable buildings. *Renew. Sustain. Energy Rev.* **2014**, *34*, 409–429. [CrossRef]
29. Elkazaz, M.; Sumner, M.; Pholboon, S.; Thomas, D. Microgrid Energy Management Using a Two Stage Rolling Horizon Technique for Controlling an Energy Storage System. In Proceedings of the IEEE 7th International Conference on Renewable Energy Research and Applications (ICRERA), Paris, France, 14–17 October 2018; pp. 324–329.
30. Vielma, J.P. Mixed integer linear programming formulation techniques. *SIAM Rev.* **2015**, *57*, 3–57. [CrossRef]
31. Elkazaz, M.; Sumner, M.; Thomas, D. Energy management system for hybrid PV-wind-battery microgrid using convex programming, model predictive and rolling horizon predictive control with experimental validation. *Int. J. Electr. Power Energy Syst.* **2020**, *115*, 105483. [CrossRef]
32. IEEE. *IEEE Guide for the Characterization and Evaluation of Lithium-Based Batteries in Stationary Applications*; No. 1679.1-2017; IEEE Std: Piscataway, NJ, USA, 2018. [CrossRef]
33. Terlouw, T.; Zhang, X.; Bauer, C.; Alskaif, T. Towards the determination of metal criticality in home-based battery systems using a Life Cycle Assessment approach. *J. Clean. Prod.* **2019**, *221*, 667–677. [CrossRef]

34. Klingler, A.-L.; Teichtmann, L. Impacts of a forecast-based operation strategy for grid-connected PV storage systems on profitability and the energy system. *Solar Energy* **2017**, *158*, 861–868. [CrossRef]
35. Richardson, I.; Thomson, M. Domestic Electricity Demand Model-simulation Example. 2010. Available online: https://repository.lboro.ac.uk/articles/Domestic_electricity_demand_model_-_simulation_example/9512927 (accessed on 1 July 2020).
36. UK Power. Average Gas and Electric Usage for UK Households. Available online: https://www.ukpower.co.uk/home_energy/average-household-gas-and-electricity-usage (accessed on 1 June 2020).
37. PVOutput. Generation Profiles for a 3.8 kW PV Station Located in Nottingham. Available online: https://pvoutput.org/ (accessed on 1 June 2020).
38. RobinHood Enegy, Economy 7 Tariff Information. Available online: https://join.robinhoodenergy.co.uk/tariffs (accessed on 1 June 2020).
39. Money Saving Expert. Time-of-Day Purchasing Tariff in UK. Available online: https://www.moneysavingexpert.com/news/2017/01/green-energy-launches-time-of-day-tariff---electricity-savings-available-but-gas-remains-pricey/ (accessed on 1 June 2020).
40. ELEXON Ltd., UK, The New Electricity Trading Arrangements for the Imbalance Market in UK 2014. Available online: https://www.bmreports.com/bmrs/?q=balancing/systemsellbuyprices/historic (accessed on 1 June 2020).
41. Ofgem. Feed-In Tariff (FIT) Rates in UK. Available online: https://www.ofgem.gov.uk/environmental-programmes/fit/fit-tariff-rates (accessed on 1 June 2020).
42. Elkazaz, M.; Sumner, M.; Thomas, D. Sizing Community Energy Storage Systems—Used for Bill Management compared to Use in Capacity and Firm Frequency Response Markets. In Proceedings of the 2020 IEEE Power & Energy Society Innovative Smart Grid Technologies Conference (ISGT), Washington, DC, USA, 17–20 February 2020; 2020; pp. 1–6.
43. BloombergNEF. A Behind the Scenes Take on Lithium-Ion Battery Prices. Available online: https://about.bnef.com/blog/behind-scenes-take-lithium-ion-battery-prices/ (accessed on 1 June 2020).
44. CCL. BYD B-Box HV—Lithium Battery Pack. Available online: https://www.cclcomponents.com/byd-b-box-high-voltage-6-4kwh-lithium-battery (accessed on 1 June 2020).
45. SMA. Sunny Boy Storage 2.5 Power Inverter. Available online: https://www.anh-technologies.co.za/pdf/sma_sunny_boy_15-25.pdf (accessed on 1 June 2020).
46. Shiffeld, t.U.o. PV Forecast Service. Available online: https://www.solar.sheffield.ac.uk/pvforecast/ (accessed on 1 June 2020).
47. F-Grid Europe. SMA Integrated Storage System SB 3600 Smart Energy Inverter. Available online: https://www.off-grid-europe.com/sma-integrated-storage-system-sb-3600-smart-energy-inverter?gclid=CjwKCAjw5fzrBRASEiwAD2OSV8QTkNn5Es4NLxChBekDTGV1-Qt1HfModDjrFphFK49LY05ftgj1cRoCWhkQAvD_BwE (accessed on 1 June 2020).
48. CALTEST Instrumnts Ltd. ZSAC Series—AC Loads—Hoecherl & Hackl. Available online: https://www.caltest.co.uk/product/zsac-series/ (accessed on 1 June 2020).
49. National Instruments. CompactRIO Systems. Available online: https://www.ni.com/en-gb/shop/compactrio.html (accessed on 1 June 2020).
50. Smart Process & Control Ltd. Smartrail X835-MID DIN Rail Multifunction Power Meter. Available online: http://www.smartprocess.co.uk/PDF/Smart-Process_SMARTRAIL-X835-MID_Datasheet.pdf (accessed on 1 June 2020).
51. Elkazaz, M.; Sumner, M.; Thomas, D. Real-Time Energy Management for a Small Scale PV-Battery Microgrid: Modeling, Design, and Experimental Verification. *Energies* **2019**, *12*, 2712. [CrossRef]

energies

MDPI

Article

Investigation of Black-Starting and Islanding Capabilities of a Battery Energy Storage System Supplying a Microgrid Consisting of Wind Turbines, Impedance- and Motor-Loads

Jürgen Marchgraber * and Wolfgang Gawlik

Institute of Energy Systems and Electrical Drives, TU Wien, 1040 Vienna, Austria; gawlik@ea.tuwien.ac.at
* Correspondence: marchgraber@ea.tuwien.ac.at

Received: 30 August 2020; Accepted: 22 September 2020; Published: 5 October 2020

Abstract: Microgrids are small scale electrical power systems that comprise distributed energy resources (DER), loads, and storage devices. The integration of DER into the electrical power system basically allows the clustering of small parts of the main grid into Microgrids. Due to the increasing amount of renewable energy, which is integrated into the main grid, high power fluctuations are expected to become common in the next years. This carries the risk of blackouts to be also more likely in the future. Microgrids hold the potential of increasing reliability of supply, since they are capable of providing a backup supply during a blackout of the main grid. This paper investigates the black-starting and islanding capabilities of a battery energy storage system (BESS) in order to provide a possible backup supply for a small part of the main grid. Based on field tests in a real Microgrid, the backup supply of a residential medium voltage grid is tested. Whereas local wind turbines within this grid section are integrated into this Microgrid during the field test, the supply of households is reproduced by artificial loads consisting of impedance- and motor loads, since a supply of real households carries a high risk of safety issues and open questions regarding legal responsibility. To operate other DER during the island operation of such a Microgrid, control mechanisms have to ensure the power capabilities and energy reserves of the BESS to be respected. Since the operation during a backup supply of such a Microgrid requires a simple implementation, this paper presents a simple master–slave control approach, which influences the power output of other DER based on frequency characteristics without the need for further communication. Besides the operation of other DER, the capability to handle load changes during island operation while ensuring acceptable power quality is crucial for such a Microgrid. With the help of artificial loads, significant load changes of the residential grid section are reproduced and their influence on power quality is investigated during the field tests. Besides these load changes, the implementation and behavior of the master–slave control approach presented in this paper is tested. To prepare these field tests, simulations in MATLAB/SIMULINK are performed to select appropriate sizes for the artificial loads and to estimate the expected behavior during the field tests. The field tests prove that a backup supply of a grid section during a blackout of the main grid by a BESS is possible. By creating the possibility of operating other DER during this backup supply, based on the master–slave control approach presented in this paper, the maximum duration for this backup supply can be increased.

Keywords: microgrid; black-start; islanding; master–slave control approach; cold load pickup

1. Introduction

In order to mitigate global warming, the integration of renewable energy will increase dramatically in the coming years. Several studies calculate a share of 50% of renewable energy in the power sector in 2030 [1,2]. A high percentage of this share will be covered by variable renewable energy, namely solar photovoltaic- and wind power systems. Due to the variable infeed-behavior of these technologies, severe power fluctuations of the total power infeed in the electric power grid are expected to become common. These power fluctuations lead to high differences between load and generation. The need for additional flexibilities in the power system to balance these differences, therefore, is indisputable. Storage technologies are one way to create additional flexibility. Long- and short-term storage technologies, therefore, are expected to be increasingly integrated in the power system [3]. Battery Energy Storage Systems (BESS) are expected to cover a part of the short-term storage demand for durations between minutes, hours or days. Examples of how the flexibility of BESS will be used to handle power fluctuations, for example, are the application of "peak-shaving" to avoid overload of transmission lines or their contribution to frequency control by providing frequency reserves. There are more of such examples, but they all have one property in common, which is that no continuous activation will take place. For example, peak-shaving may only be necessary in times of high stress of the grid, otherwise the surplus power may be transmitted to places of higher demand. The corresponding flexibility in the latter situation would be transmission of power, leading to the necessity of grid expansion, which is also listed as a requirement for flexibility in the future power system in several studies [1,2]. Because BESS are considered not to perform peak-shaving, frequency control, or other services, continuously, the idea of "value-stacking" arises. Other services or applications, which are not directly related to handle power fluctuations in the future power system, are considered as a way to increase the profitability of a BESS, by providing them besides other services. Islanding can be considered as one of such applications. Due to the high amount of renewable energy in the future power system, the risk of blackouts increases. By building local Microgrids and perform an island operation during such blackouts, the reliability of supply can be increased.

The increasing amount of decentralized energy resources (DER) offers the possibility to build such Microgrids (MG). According to Marney et al. [4], Microgrids comprise low voltage distribution systems with distributed energy sources, storage devices, and controllable loads, operated connected to the main power grid or islanded, in a controlled, coordinated way. According to Mahmoud et al. [5], there are several ways to to categorize MG, for example, by their application type (Utility MG, Commercial MG, Industrial MG, remote/off-grid MG), their system structure (single-stage/two-stage power-conversion systems), the type of control (centralized, decentralized or ditributed), etc. Based on the application type, the MG are embedded into the low- or medium-voltage level of the main grid, depending on the size of the integrated loads and DER. Remote MG are an exception, which are operating without any connection to the main grid. A Microgrid has two operation modes, which are termed as "grid-connected mode" and "island mode". The operation of a Microgrid separate from the main grid in island mode is based on suitable control algorithms. In the case of a blackout of the main grid, a Microgrid is able to switch from grid-connected mode to island mode, either via a black-start or by a smooth transition. Therefore, an appropriately designed Microgrid improves reliability and resiliency by providing a back-up system against grid faults [6]. There may also be an application for Microgrids to facilitate the restoration process after blackouts [7]. Besides such applications of Microgrids which are "embedded" into the main grid, the most straightforward application of Microgrids is their application in remote areas, where it is not possible or economically not feasible to provide a sufficient power link to the main grid, as, for example, described in [8].

Many DER are connected to the grid via converters. To be able to operate an islanded Microgrid, at least one of these DER must have island capabilities in order to control the voltage and frequency in the Microgrid. Such converters are called "grid-forming" converters. The major difference between grid-forming converters and conventional "grid-following" converters is their ability to build the grid angle on their own, whereas grid-following converters rely on the determination of the grid angle based on the voltage measurement at their point of common coupling (PCC). The supply of an islanded Microgrid by multiple grid-forming units requires the application of appropriate load sharing techniques. Conventionally, a voltage- and frequency droop control is adopted in such cases, as, for example, described by Katiraei and Iaravni [9] or by Barklund et al. [10]. However, conventional droop control neglects the influence of the coupling of the grid-forming units [11]. This results in large power circulating-currents and power fluctuations of the Microgrid [12]. Improved droop control strategies, as, for example, proposed by Peng et al. [11], are using virtual impedances in order to take into account the coupling of the grid-forming units. For the application of a Microgrid, to build a backup supply during the blackout of the main grid, as described above, the use of several grid-forming units may be not very practical, at least for smaller Microgrids. Most of the DER, which will be integrated into the main grid in the future are based on variable renewable generation. Therefore, they are not fully capable of contributing to power balancing during a backup supply in an islanded Microgrid because they are connected via grid-following converters and may only cover a share of the energy demand of the loads in an intermittent way. However, as described above, the integration of decentralized BESS nearby to larger DER based on variable renewable energy in the future is very likely in order to handle the power fluctuations of these DER. By designing such BESS to be able to switch between a grid-following and a grid-forming mode, this opens the possibility to build Microgrids of certain sections of the main grid. The most practicable approach for such an arrangement is a master–slave control approach. Such a master–slave approach is characterized by using only one grid-forming unit, which represents the "master" during island operation, while the other DER are acting as "slave" in a grid-following mode. This approach is usually also used when Microgrids are supplied with large diesel generators, for example as described in [13], but can also be used to achieve a sustainable supply of Microgrids based only on renewable energy. Although master–slave control approaches for Microgrids have already been investigated in several studies [13–16], the major drawbacks can be identified by the need for communication [14–16] and the lack of experimental investigations in real world grid sections, since all studies are limited to simulations. An example for real world measurements can be found in [13] where power quality measurements in an artificial test Microgrid were performed. However, Garde et al. [13] conclude the lack of energy management for the BESS, which acts as a master unit in their test Microgrid. This paper presents a master–slave approach without the need for communication and the possibility to perform a simple energy management of the BESS.

As already mentioned above, BESS are considered to become a necessity to provide flexibility in the future power system as short-term storages. Due to decreasing prices of Li-Ion battery cells [17], this technology is considered to make up the majority of BESS which are used in the future. Against the background of value-stacking, islanding is considered as an additional service BESS can provide to increase their profitability, while at the same time increasing the reliability of supply in the case of blackouts of the interconnected power grid by building local Microgrids. Assuming such Microgrids to comprise a high share of variable renewable energy, an easy approach for ensuring their functionality during island operation is desired in order to lower the requirements of the energy reserves of the BESS and to increase the maximum duration of island operation. This paper presents a master–slave approach based on a variable frequency, which allows the power infeed of variable renewable energy during island operation of a Microgrid, without the need for additional communication. Furthermore, the black-start and island operation of a real Microgrid is investigated, within which this master–slave approach is applied. This real Microgrid comprises a BESS, wind turbines, and loads, which are used to reflect the behavior of a residential grid section. The motivation for this paper is to investigate the possibility of black-starting and operating a residential grid section during an assumed blackout of the interconnected power grid that the residential network is connected with. Whereas the BESS and the wind turbines are integrated into the corresponding grid section during normal operation, the connection of the nearby residential grid section, comprising several households, was not possible due to safety reasons. Therefore, several impedance- and motor-loads are used to reproduce significant load-steps of this residential grid section. Via field tests, the black-starting and islanding capabilities of the Microgrid and the master–slave approach presented in this paper are investigated. For preparation of these field tests, simulations in MATLAB/SIMULINK were performed, which allowed an evaluation of the expected results.

To present the corresponding results and findings, this paper is structured as follows. Section 2 presents the simulation model used in the simulations and presents the simulation results of a black-start and during load-changes in island operation. Section 3 presents the measurement results which were taken during field tests in a real Microgrid. Section 4 concludes the paper.

This paper focuses on Microgrids that use a BESS as master unit in a master–slave control strategy described above. Related to these types of Microgrids, the following Section 1.2 discusses more details on the black-start of such Microgrids and Section 1.3 discusses more details on master–slave control approach which is investigated in this paper.

1.1. Nomenclature

To tell apart unit-based and normalized values, uppercase and lowercase letters are used. Uppercase letters are related to unit-based values, while lowercase letters are related to normalized values. Normalized values are using the unit "pu". To tell apart phasors and instantaneous values, the index "t" is used for instantaneous values. Bold symbols identify a vector. Nominal values are identified with the index "n". A list of all symbols used in this paper is summarized in Appendix A.

1.2. Black-Start

The black-start of a Microgrid defines the process of restoring it after a shutdown without any external voltage support. The operation of a Microgrid is often preceded by a black-start, which may be necessary in the case of a blackout of the main grid, when no smooth transition of the Microgrid from grid-connected into island mode takes place, or in case of a remote Microgrid, where a supply may only be given during specific time periods during the day. The black-start of a Microgrid is a challenging task to which special attention must be paid. The main issue with black-starting an island grid is a phenomenon which is termed as "cold load pickup". In particular, a high penetration of thermostatically controlled devices leads to a high power demand during the black-start of a Microgrid [18]. During normal operation, their activity is evenly distributed, whereas during a black-start all of these devices drain power at the same time, as the thermostatically controlled area got "cold" during the blackout. The same behavior occurs with many devices, as their random switching during normal operation may be lost during a black-start. In addition, their power consumption during start-up may be higher than during normal operation, for example, because of certain start-up routines. In addition, most types of loads drain power depending on the temperature. If they possess a negative temperature coefficient, their resistance decreases with rising temperature. Therefore, during a start-up process of a load where the load temperatures are lowest, they drain higher power than in their nominal operating point. The remaining equipment of a Microgrid, as for example transformers and cables, also have a higher power demand during start-up. The inrush currents of transformers and the initial charging process of the capacitances of cables are the main reasons for this higher power demand. The challenge of performing a black-start, therefore, is to handle the "cold load pickup" in terms of both active- and reactive power.

The most important property to perform a black-start is the black-start ability of at least one grid-forming unit in the Microgrid. Assuming a BESS to be this grid-forming unit, such a black-start ability can only be achieved if all of its components and control systems can be activated and maintained without any external power supply. The most common solution to reach such a condition is the use of an additional uninterruptible power source (UPS) to supply these components of the BESS. The primary battery of the BESS itself can also be an option to serve as a power supply. However, as some of the components are necessary to start-up and run the battery itself and the AC-output voltage of the BESS often differs from the voltage that is needed to supply all components and auxiliary systems (e.g., 400 V), the use of an additional UPS is also a common approach. One of the biggest consumer of all necessary components to maintain a BESS is its heating, ventilation, and air conditioning (HVAC) system. Therefore, the UPS must be designed to handle such a power consumption or the BESS must be designed to start-up without an active HVAC system.

In general, black-start ability also requires the BESS to be operable automatically, remotely or by local staff. It has to be guaranteed that all breakers are functional without any external power supply in order to start-up the BESS and establish a connection to the Microgrid. A corresponding start-up routine has to guarantee a procedure to appropriately control these breakers in an intended sequence. A crucial point of such a start-up routine is an appropriate island detection. The island detection guarantees that a black-start and further connection to a Microgrid, by closing the relevant breaker at the PCC of the BESS, is only possible when the Microgrid is without voltage and not connected to the main grid. Otherwise, massive damage may be caused to the BESS and other components due to very high currents as a result of possible phase differences between the voltage of the BESS and the voltage of the main grid. Therefore, a synchronization unit is essential for a connection to the main grid. At the end of a start-up routine, the relevant breaker of the PCC is closed in order to energize the Microgrid. As already mentioned above, the BESS has to be capable of handling the cold load pickup that may arise during this process. A simple way to deal with the cold load pickup is to gradually increase the system voltage instead of performing a black-start with nominal voltage during this last step of connecting to the Microgrid. In addition, the inrush currents of all transformers and the charging currents of the cable capacitances in the island grid can be decreased by doing so. Although most of the electrical equipment and loads will not be damaged due to operation with under-voltage, there may be processes that rely on voltage magnitudes within certain limits. For loads that are part of such processes, a corresponding breaker logic has to guarantee that their connection is only established after a certain voltage limit has been reached. In addition, protection devices may have to be reconfigured to handle the gradual increase of the system voltage. In addition, the protection devices of the BESS that are configured to trip as a consequence of under-voltage have to be reconfigured.

1.3. Island Operation

The black-start of a Microgrid is followed by the island operation. The challenge during island operation is to manage the active- and reactive power demand that results from the load behavior, and the active- and reactive infeed that results from DER in the Microgrid. Therefore, an islanded Microgrid that contains several DER requires suitable control mechanisms to ensure that the maximum power of the grid-forming unit is respected. The implementation of such control mechanisms offers the possibility of operating an islanded Microgrid, the peak load, and the peak infeed that exceed the nominal power of its grid-forming unit. At least a part of the load can be supplied by DER while the grid-forming unit only has to deliver balancing power. The challenge of such an operation is the impact of a sudden imbalance between load and generation. The control mechanisms mentioned above have to ensure that such imbalances become controllable by the grid-forming unit.

The ability of grid-forming converters to generate the grid angle on their own offers the potential of dictating the frequency to actively influence the power output of other generating units and the power consumption of loads in an islanded Microgrid via pre-defined frequency characteristics.

The simplest approach for such control mechanisms is using load-shedding for a surplus of load and power reduction for a surplus of infeed, in case either of them exceeds the power limits of the grid-forming unit. The frequency in the Microgrid can be used as control signal for both operations without the need for additional communication. For load-shedding, additional breakers that trip at certain frequency limits and a stepwise load-shedding plan with prioritization of loads may be necessary. For power reduction of generation units, the "limited frequency sensitive mode" (LFSMO) can be used. According to latest grid codes [19], generation units have to reduce their power output in case of a frequency that exceeds a limit of 50.2 Hz. This mechanism can be used in an islanded Microgrid to actively influence the power output of DER through the LFSMO. This kind of frequency control approach can not only be used to limit power flows beyond the limits of the BESS, but can also be used to perform an energy management to manage the state of charge (SoC). Such a frequency control approach is already being used in home storage systems [20]. A corresponding frequency characteristic to be used in a Microgrid described above is shown in Figure 1.

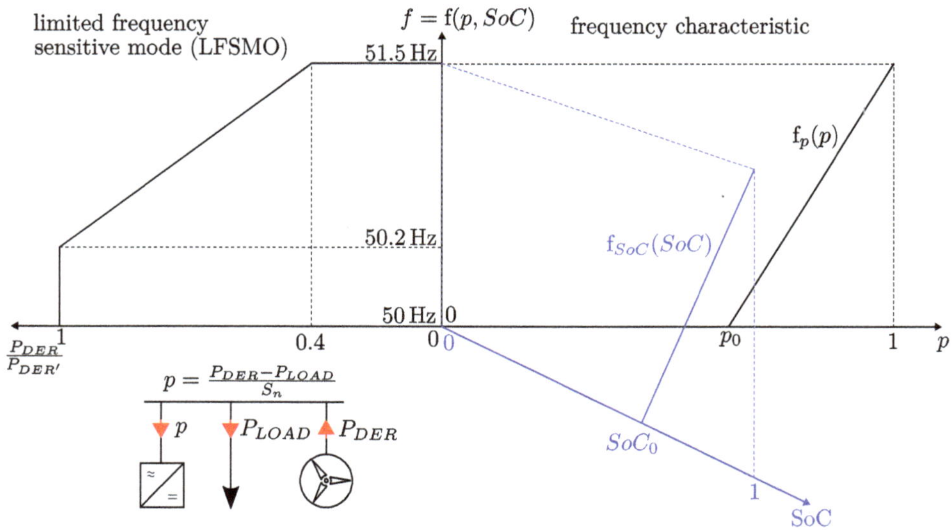

Figure 1. Characteristic to control the frequency of a grid-forming BESS in order to reduce the currently possible power output P'_{DER} of other generating units to a reduced power output P_{DER} in case of a high SoC or in case the imbalance between generation and load p reaches values near 100% (rated power of the BESS).

The normalized power intake of the grid-forming BESS is defined by $p = \frac{P_{DER} - P_{LOAD}}{S_n}$ and equals the power imbalance between the power infeed of DER P_{DER} and the power consumption of loads P_{LOAD} in the Microgrid, which is normalized to the nominal apparent power S_n of the BESS. The values p_0 and SoC_0 can be used to parameterize the frequency characteristic. The grid-forming BESS controls its output frequency according to the resulting frequency characteristic. As a consequence of the LFSMO, the DER starts to reduce their currently possible power output P'_{DER} above a frequency of 50.2 Hz in 50 Hz-systems to a reduced power output P_{DER}. According to Figure 1, the frequency f of the BESS is a function of the normalized power intake p and the state of charge SoC, defined by $f = \mathrm{f}(p, SoC)$. To deal with situations when $p > p_0$ and $SoC > SoC_0$, this function is defined by

$$f = \mathrm{f}(p, SoC) = \max[\mathrm{f}_p(p), \mathrm{f}_{SoC}(SoC)]. \quad (1)$$

The frequency curve may be expanded by an additional part to reduce the output frequency of the grid-forming BESS in order to trigger corresponding under-frequency relays that initiate load-shedding as mentioned above.

During island operation, the massively reduced short-circuit power leads to larger voltage excursions during imbalances of load and generation. In addition, during island operation, the power quality still remains an important issue.

2. Simulation

For the preparation of the field tests, a converter model in MATLAB/SIMULINK was created in order to investigate the expected behavior during island operation. The estimation of voltage dips during load-steps and start-up of motors was a goal to choose permissible sizes for the load and the motors. Furthermore, it was aimed to investigate the influence of control parameters on these voltage dips in order to get information under which conditions instabilities and maximum voltage dips during the field tests can be expected.

2.1. Model Description

The ability of a BESS to black-start and maintain a Microgrid relies on an appropriate control design of its converter, which has already been named grid-forming converter above. As many BESS not only operate in island mode but also in grid-connected mode, the control design of theirs converter has to be able to switch between these two operation modes. Based on a simplified converter model, Figure 2 shows an appropriate control design which is capable of operating in both "island mode" and "grid-connected mode".

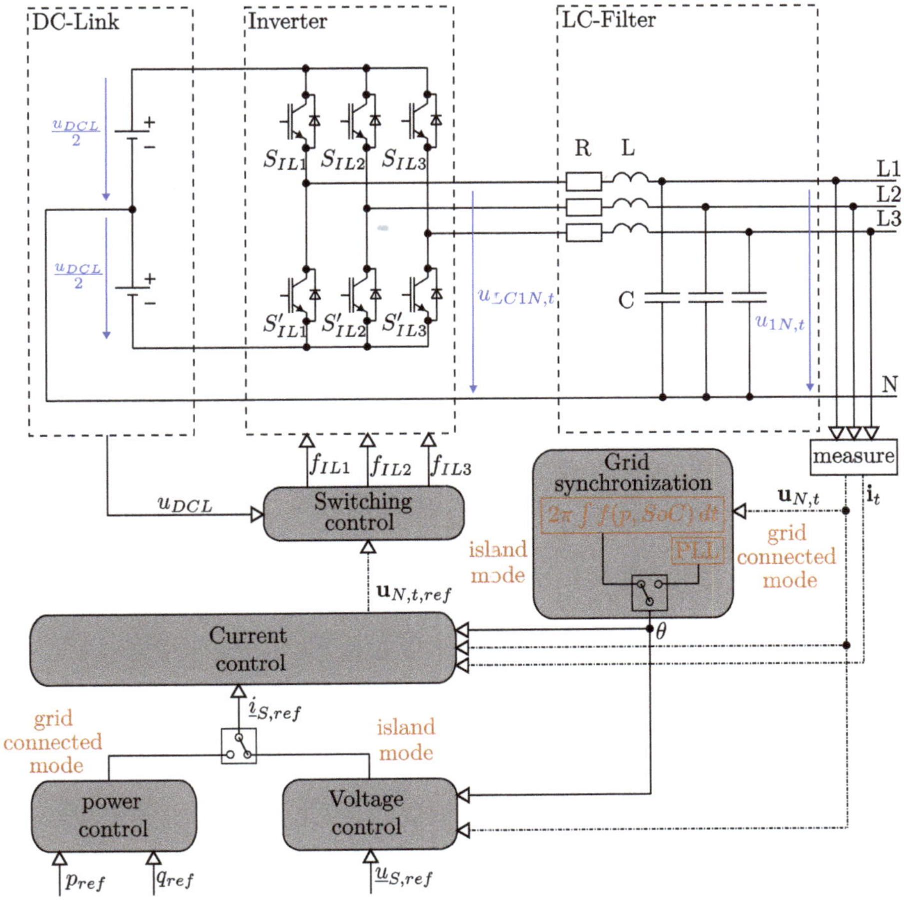

Figure 2. Converter model and its control to realize both island mode and grid-connected mode. The blue color indicates normalized voltages. The orange color indicates selectable control elements during grid-connected and island mode.

The simplified converter model of a three-phase voltage source converter (VSC) shown in Figure 2 is a two-level converter, which consists of a DC-Link, an inverter, and an LC-filter. For simplification, the DC-link is modeled as a constant voltage source. The inverter model consists of a self-commutated three-phase bridge, which is controlled by a pulse-width modulation signal. The resistance in the LC-filter at the output of the converter represents the parasitic resistance of the inductance.

The lower part of Figure 2 shows the control structure of the converter model. It consists of a "grid synchronization", which builds the grid angle θ. This grid angle is used in the "current control" to generate a reference voltage $\mathbf{u}_{N,t,ref}$ based on a current reference space vector $\underline{i}_{S,ref}$. The control structure includes two switches by which the operation mode, either grid-connected or island mode, and therefore a corresponding converter control structure is selected. In grid-connected mode, the current reference signal is generated by the "power control", which calculates $\underline{i}_{S,ref}$ based on active- and reactive power reference signals p_{ref} and q_{ref}. The grid angle θ during grid-connected mode is delivered via a synchronous reference frame phase-locked loop (SRF-PLL). During island mode, the grid angle θ is delivered based on a parametrizable value of the angular frequency ω with $\theta = \int \omega \, dt$. The current reference space vector during island mode is generated by the "voltage control", which uses a voltage reference space vector $\underline{u}_{S,ref}$ to calculate it. The current control is followed by the "switching control" which generates the switching signals f_{IL1}, f_{IL2} and f_{IL3} based on the voltage reference signal $\mathbf{u}_{N,t,ref}$ and the DC-link voltage u_{DCL} in order to control the switches S_{IL1}, S_{IL2} and S_{IL3} of the inverter. The complementary switching signals f'_{IL1}, f'_{IL2} and f'_{IL3} are used to control the switches S'_{IL1}, S'_{IL2} and S'_{IL3} of the inverter. More details about the control structure during grid-connected mode can be found in [21,22]. For the simulations in this paper, only the control structure during island mode is relevant.

In island mode, the voltage- and current control builds a cascaded control structure, which is presented in more detail in Figure 3.

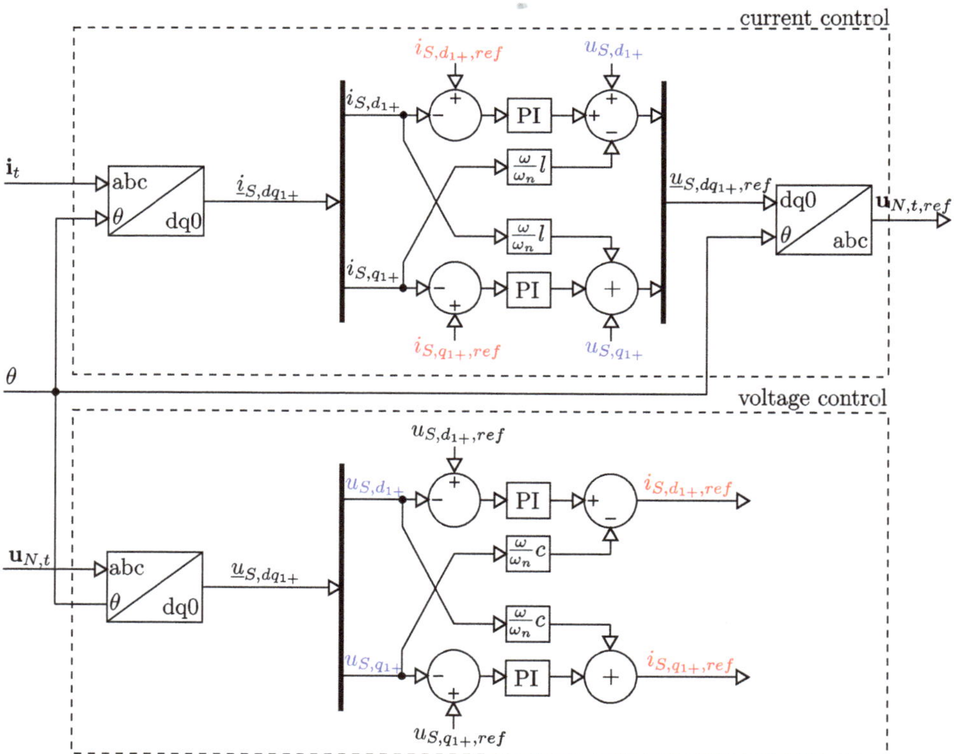

Figure 3. Cascaded voltage- and current control of Figure 2 in island mode.

Based on the grid angle θ, the measured voltage- and current vectors $\mathbf{u}_{N,t}$ and $\mathbf{i}_t$ are used to calculate corresponding space vectors $\underline{u}_{S,dq_{1+}}$ and $\underline{i}_{S,dq_{1+}}$ via Park-transformation. In the voltage control, the voltage space vector $\underline{u}_{S,dq_{1+}}$ is used to control the corresponding d- and q-components via PI-control based on a reference space vector $\underline{u}_{S,dq_{1+},ref}$. The coupling between the d- and q-components due to the LC-filter is taken into account by corresponding decoupling terms $\frac{\omega}{\omega_n}c$, where c is the normalized capacitance of the LC-filter, which is defined by $c = \frac{\omega_n C S_n}{U_n^2}$, with S_n as the nominal apparent power and U_n as the nominal voltage of the converter. The resulting current reference space vector $\underline{i}_{S,dq_{1+},ref}$ of the voltage control is used as input of the current control, which uses the current space vector $\underline{i}_{S,dq_{1+}}$ to control the corresponding d- and q-components via PI-control based on this current reference space vector. The coupling between the d- and q-components due to the LC-filter is taken into account by corresponding decoupling terms $\frac{\omega}{\omega_n}l$, where l is the normalized inductance of the LC-filter, which is defined by $l = \frac{\omega_n L S_n}{U_n^2}$.

The tuning of the PI-controllers of the voltage- and current control can be conducted based on a stepwise approach according to Henninger [23], where the first step is the definition of a freely choosable time constant τ_i according to Yazdani and Iravani [24], which allows the current control to be represented by a first-order filter with the time constant τ_i. The voltage control can then be tuned by using the symmetrical optimum according to Yazdani and Iravani [24]. The resulting parameters $K_{U,P}$ and $K_{U,I}$ of the PI-controllers of the voltage control and the parameters $K_{I,P}$ and $K_{I,I}$ of the PI-controllers of the current control based on this tuning approach result in the following definitions:

$$K_{I,P} = \frac{L}{\tau_i} \tag{2}$$

$$K_{I,I} = \frac{R}{\tau_i} \tag{3}$$

$$K_{U,P} = \frac{C}{\tau_i}\sqrt{\frac{1-\sin\Phi_R}{1+\sin\Phi_R}} \tag{4}$$

$$K_{U,I} = \frac{K_{U,P}}{\tau_i}\frac{1-\sin\Phi_R}{1+\sin\Phi_R} \tag{5}$$

with Φ_R as the phase margin of the voltage control. A detailed description of the tuning of the PI-controllers of the voltage- and current control and a corresponding derivation of the parameters is given in [22]. Assuming the parameters of the LC-filter to be given, the tuning of the voltage- and current control is only dependent on the time constant τ_i and the phase margin Φ_R.

Figure 4 shows the single line diagram of the Microgrid which is investigated in the simulation. This Microgrid is based on a real Microgrid which was also investigated in [25]. In an extended version of this Microgrid, field tests were also performed, which are presented in Section 3. It consists of a BESS based on Li-Ion technology with a rated power of 2.6 MVA, which is connected via line 1 to a 30 kV-substation. Via line 2 and a transformer, an active load and an asynchronous machine are connected to the substation.

As the Microgrid shown in Figure 4 consists of four battery converters, which are connected in parallel, the converter model shown in Figure 2 has to be adapted in order to consider this parallel connection. In case of a parallel connection of several converters, an equivalent converter model can be used by adapting the values of the LC-filter at the output of the model accordingly. Every single converter during a parallel operation behaves identically compared to a single operation. The only difference is the higher apparent power and the changed impedance due to parallel connection.

Assuming the LC-filter of a single converter to be L', R' and C', the resulting values of the LC-filter of the equivalent converter model can be calculated by

$$L = \frac{L'}{n} \tag{6}$$

$$R = \frac{R'}{n} \tag{7}$$

and

$$C = n \cdot C' \tag{8}$$

with n as the number of parallel connected converters. As the cascaded voltage- and current control shown in Figure 3 uses normalized values of the voltage $\mathbf{u}_{N,t}$ and the current $\mathbf{i}_t$ as input, the parameters of the PI-controllers for the equivalent converter model remain identical compared to the single converter model. The higher apparent power of the equivalent converter model

$$S_n = n \cdot S_n' \tag{9}$$

only reflects in the normalization of the measured voltage $\mathbf{U}_{N,t}$ and current $\mathbf{I}_t$. Therefore, the control dynamics of the equivalent converter model is identical to the single converter model. Only the normalization and the LC-filter has to be adapted in the equivalent converter model compared to the single converter model.

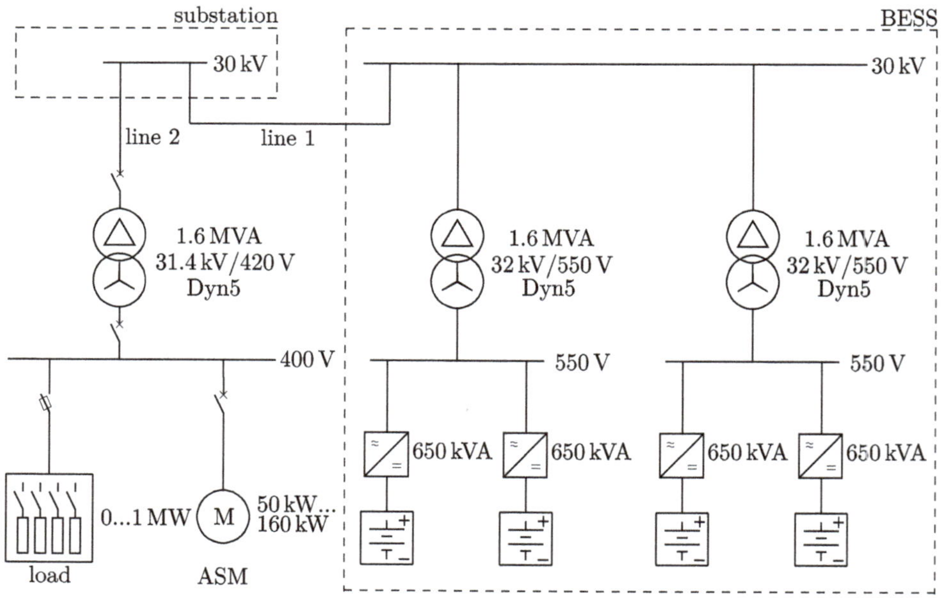

Figure 4. Single line diagram of the island grid investigated in the simulation.

The relevant parameters of the components of the Microgrid shown in Figure 4 are summarized in Table 1.

Table 1. Parameters of the components in Figure 4 used in the simulation.

Characteristic	Value
line 1	
R'_{l1}	$0.26\,\Omega\,\text{km}^{-1}$
L'_{l1}	$0.43\,\text{mH}\,\text{km}^{-1}$
length	50 m
line 2	
R'_{l2}	$0.7\,\Omega\,\text{km}^{-1}$
L'_{l2}	$0.5\,\text{mH}\,\text{km}^{-1}$
length	40 m
transformers (values used for all transformers)	
u_k	0.06 pu
P_K	13 kW, i.e., 0.0082 pu
equivalent converter model	
n	4
L	$L = \frac{L'}{n} = \frac{260\,\mu\text{H}}{4} = 65\,\mu\text{H}$
R	$R = \frac{R'}{n} = \frac{1\,\text{m}\Omega}{4} = 0.25\,\text{m}\Omega$
C	$C = n \cdot C' = 4 \cdot 342\,\mu\text{F} = 1368\,\mu\text{F}$
Φ_R	50°

2.2. Simulation of Load-Steps and Motor Start-Ups

As already mentioned in Section 1.3, an important property of a Microgrid is its robustness against load changes. Such load changes can be divided into active- and reactive load changes. In the simulations as well as in the field tests, the significant active- and reactive load changes of a residential grid section are reproduced by a load-bank and motor start-ups. During the island operation of a Microgrid, its robustness against such load changes may be crucial to guarantee a stable grid operation with sufficient power quality. According to Bagert et al. [26], a value of 0.8 pu can be assumed as reasonable value for a minimum dynamic voltage drop in a Microgrid. The converter model described in Section 2.1, therefore, is used to simulate the voltage drop during load-steps and the start-up of motors in order to identify the minimum dynamic voltage drop that can be expected during the field tests. Based on these simulations, the choice of the size of the loads and the motors for the field tests is made. As these voltage drops are depending on the control parameters of the converter, their influence is investigated in order to provide corresponding information for the field tests. This also applies to the motors, the successful start-up of which also depends on the control parameters.

The load bank shown in Figure 4 represents a impedance load, which is used to simulate active power load-steps. Figure 5 shows the results of the simulation of an active power load-step of 800 kW through the load bank.

Figure 5a shows the active power, Figure 5b the reactive power, Figure 5c the voltage at the PCC of the BESS, and Figure 5d the voltage at the connection point of the asynchronous machine. The different curves show the results for different values of the converter control time constant τ_i, which defines the parameters for the PI-controllers of the voltage- and current control shown in Figure 3 according to Equations (2)–(5). Other than in the main grid, where an active power load-step usually is coupled with a change in the frequency, the frequency in the islanded Microgrid is fixed at a value of $f = \frac{\omega_n}{2\pi}$ in case no frequency droop characteristic is included in the control structure shwon in Figure 2. An active power load-step in the islanded Microgrid, therefore, only leads to a voltage drop, which is to be controlled by the converter control. With an increasing value of τ_i, the voltage drop at the converter terminals increases as well. A higher time constant τ_i results in a longer settling process until the stable active power operating point of $800\,\text{kW} \equiv 0.3\,\text{pu}$ is reached. With a time constant of $\tau_i = 0.4\,\text{ms}$, the voltage drop at the converter terminals leads to a value of $u = 0.4\,\text{pu}$. Due to the influence of the transformers and lines in the Microgrid, the voltage drop at the 400 V-bus of the load bank has slower dynamics and leads to a less drastic voltage drop to a value of $u = 0.8\,\text{pu}$.

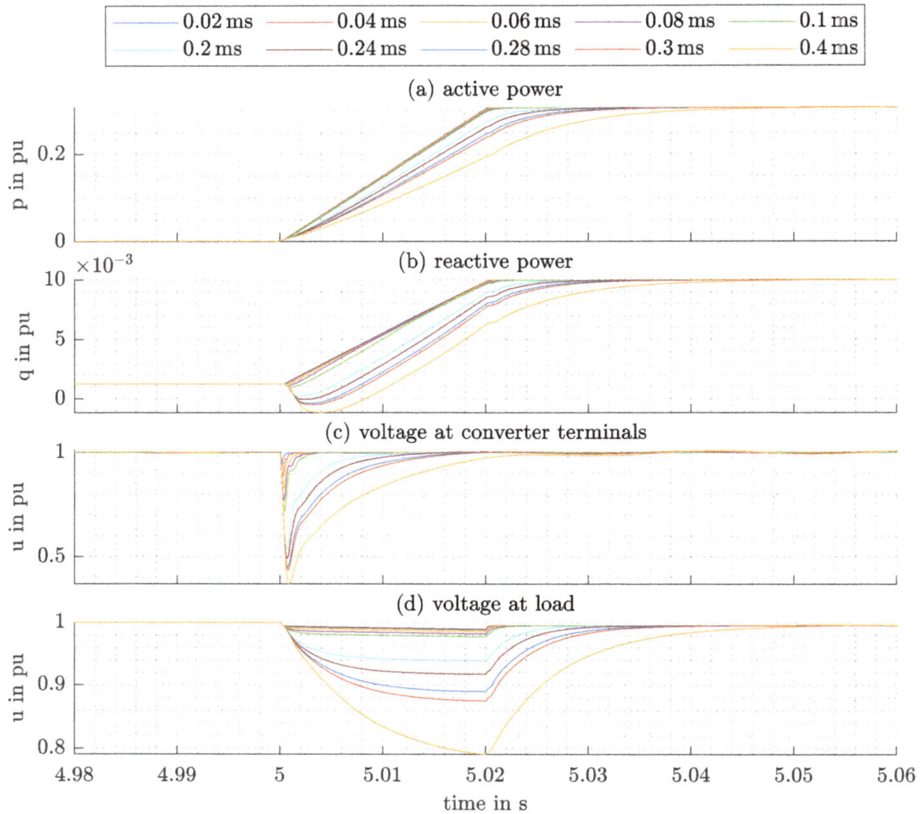

Figure 5. Results of the simulation of a load-step of 800 kW for different values of τ_i.

Figure 6 shows a sensitivity analysis of the minimum voltages, at the PCC of the converter, and the 400 V-bus of the load bank, over the settling time, until a stable operating point of the active power is reached. For this consideration, the settling time is defined as the duration between the beginning of a load-step and the moment when the voltage gradient during the voltage restoration process becomes smaller than 0.3 pu/s. The minimum voltages $min(u_C) \approx 0.4$ pu and $min(u_{load}) \approx 0.8$ pu of Figure 5 with a time constant of $\tau_i = 0.4$ ms are marked in Figure 6. Especially for low values of the time constant τ_i, the size of the voltage drop and the corresponding duration of the settling process back to a nominal value is very sensitive to changes of τ_i. Regarding power quality, the voltage at the load is relevant. The corresponding voltage behavior can be classified as a short voltage dip, which is not likely to affect the operation of connected devices in the Microgrid.

Based on the assumption that a voltage drop to 0.8 pu is considered as the minimum dynamic voltage drop according to Bagert et al. [26], a maximum load-step of 800 kW is selected to be used in the field test, taking into account that the control parameters of the real converter are parametrized faster than a corresponding $\tau_i = 0.4$ ms.

Besides the influence of an active power load-step, the influence of a reactive power load-step is also relevant during island operation. When starting an asynchronous machine, its starting current reaches values five to seven times higher than its nominal current. A motor start-up, therefore, can be used to simulate a massive reactive load-step in the Microgrid. Figure 7 shows the results of the simulation of a start-up of a 160 kW asynchronous machine to investigate such a situation of a massive reactive power load-step.

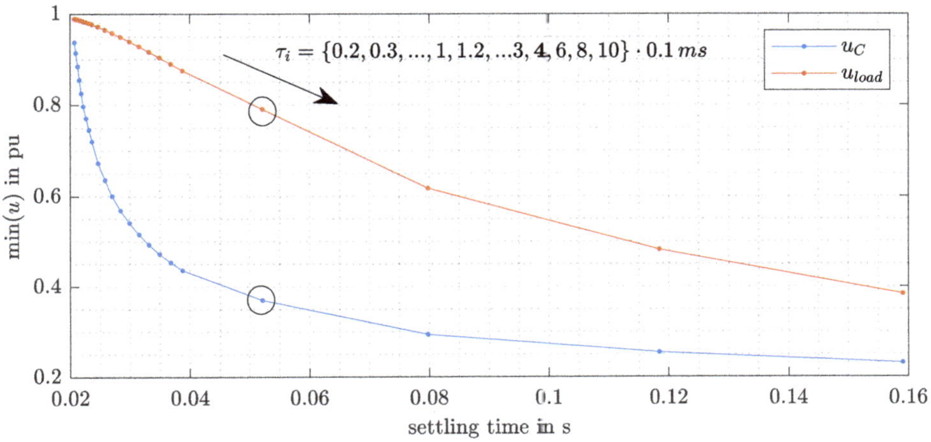

Figure 6. Sensitivity analysis of the influence of a 800 kW load-step on the voltage u_C at the PCC of the converter and the voltage u_{load} at the connection point of the load in dependence of the settling time until a stable operating point of the active power is reached. The results of $\tau_i = 0.4$ ms are circled.

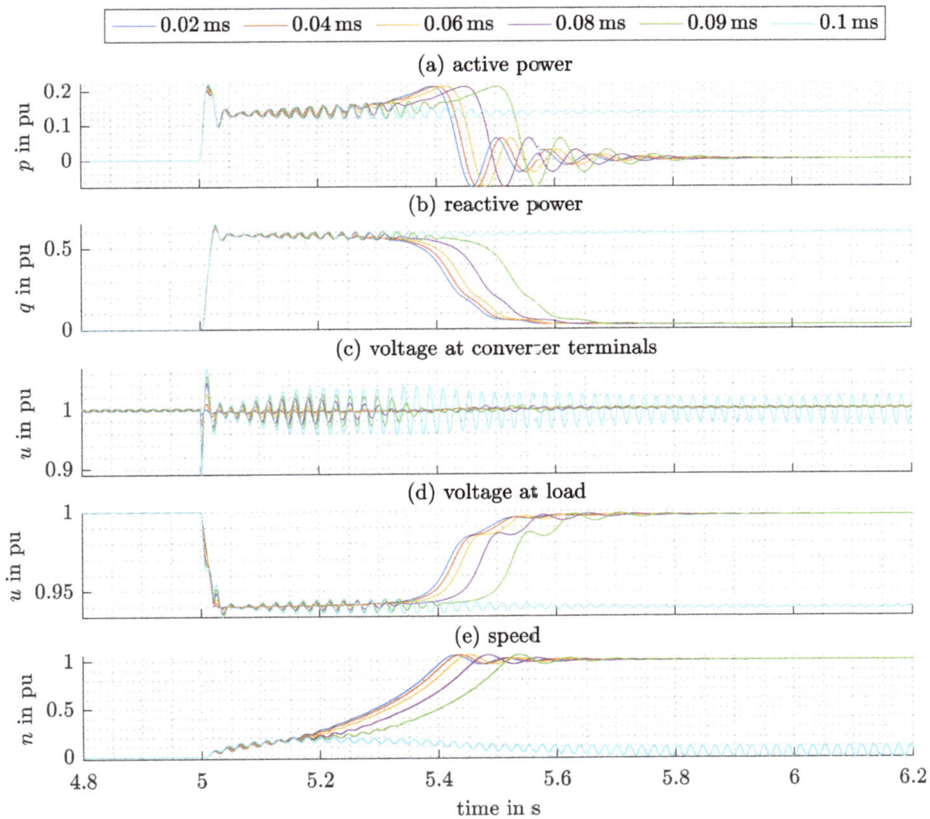

Figure 7. Results of the simulation of a motor start-up of a 160 kW asynchronous machine for different values of τ_i.

The parameters of the asynchronous machine are defined by a preset-model available in MATLAB/SIMULINK. Figure 7a shows the active power, Figure 7b the reactive power, Figure 7c the voltage at the PCC of the BESS, Figure 7d the voltage at the 400 V-bus of the asynchronous machine, and Figure 7e shows the speed of the asynchronous machine. The different curves show the results for different values of the converter control time constant τ_i. With the increasing value of τ_i, the voltage at the PCC of the converter starts to oscillate, resulting in oscillations of the active- and reactive power. With values higher than $\tau_i = 0.09$ ms, the asynchronous machine can no longer be started. The corresponding results for an unstable motor start-up with $\tau_i = 0.1$ ms are also shown in Figure 7. Figure 7e shows that the motor is not able to reach nominal speed any more. Therefore, the active- and reactive power demands remain at rather high values. This results in an ongoing voltage drop at the 400 V-bus of the motor. A converter controller time constant of $\tau_i = 0.1$ ms poses the stability limit for the start-up of a asynchronous machine with an apparent power of 160 kW in the Microgrid. Only with lower values of $\tau_i < 0.1$ ms is a motor start-up possible. Such stability limits can be determined for different motor sizes. Table 2 summarizes these stability limits.

Table 2. Summary of stability limits for different motor types.

Preset Model Motor Type	Largest Value of τ_i
37 kW	0.4 ms
75 kW	0.2 ms
110 kW	0.1 ms
160 kW	0.09 ms

Table 2 lists the largest time constant τ_i with which a start-up of the corresponding motor is still possible. The different motor types represent the available preset-models in MATLAB/SIMULINK. Only a motor size of 200 kW was available for the field tests. However, based on the results of Table 2, with a parameterization of control parameters on the real converter which are faster than a corresponding $\tau_i = 0.09$ ms, it can be assumed that a successful start-up during the field tests is possible.

3. Field Tests

During the field tests, a small residential grid section was disconnected from the main grid in order to build an islanded Microgrid. More details about the field test setup can also be found in [25,27]. The corresponding single line diagram of this Microgrid is shown in Figure 8 and consists of the BESS, which has already been described in Section 2.1, a local grid, which acts as "artificial" representation of a low voltage grid, and a wind park feeder.

The local grid consists of a freely adjustable active power load bank and an asynchronous machine, which acts as reactive power sink. This local grid is capable of imitating massive load changes of a low voltage grid. The wind park consists of several feeders, each of them serially connecting several wind turbines. Each of these wind turbines has an apparent power of 3 MVA, which is fully deployed at a wind speed of 13 m s^{-1}.

One of these feeders, which serially connects two wind turbines, is connected to the islanded Microgrid, the other feeders were disconnected during the field test. The single line diagram shown in Figure 4 which was used for simulation is identical to the single line diagram shown in Figure 8, but does not include the wind park as it was not considered in the simulation model in Section 2.1. The goals of the field test include the investigation of

- the black-start of the Microgrid,
- the synchronization of the wind park to the Microgrid,
- the island operation with an active power infeed of the wind park, where an implementation of the frequency control characteristic described in Section 1.3 based on a master–slave control approach is active and
- the investigation of active- and reactive load changes of the local grid.

The findings of the corresponding field tests are described in the following sections.

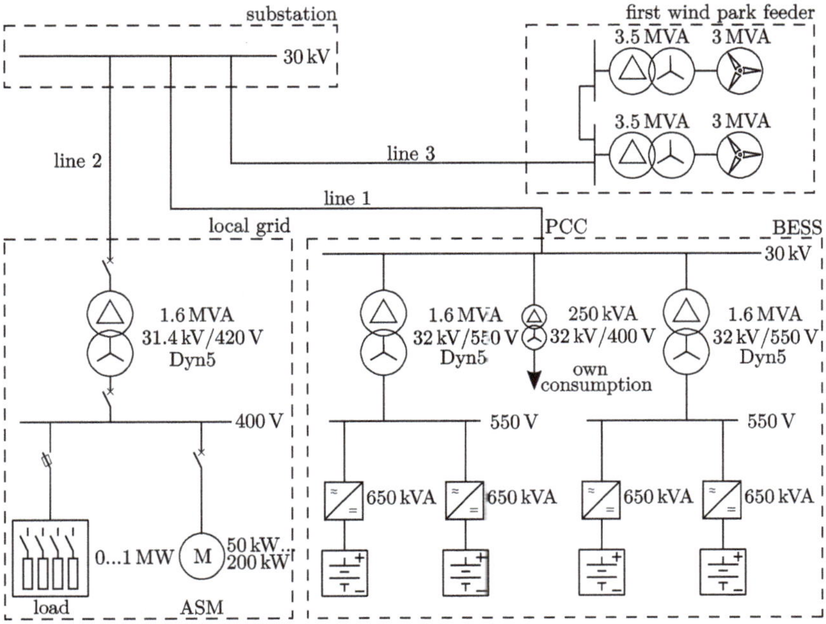

Figure 8. Single line diagram of the island grid investigated in the field tests.

3.1. *Black-Start of the Microgrid*

As already mentioned, the cold load pickup during the black-start of a Microgrid is a challenging issue. The corresponding high current demand of the cold load pickup may be misinterpreted as fault condition by the protective relays and therefore carries the risk of preventing a successful black-start as the protective relays may trip. An example for the behavior of the cold load pickup is shown in Figure 9, which presents the current demand during connection of the auxiliary transformer of the Microgrid shown in Figure 8.

Due to the nonlinear relationship of the flux and the magnetizing current of the transformer, the connection of the transformer leads to high current peaks during the first several cycles. Their magnitude is initially 6–10 times higher than the rated load current. The auxiliary transformer with a nominal power of 250 kVA has a rated load current of 4.6 A, which corresponds to a current demand of 0.1 pu for the BESS. The maximum expected inrush current corresponds to a current demand of 1 pu for the BESS. The magnitude of the inrush current shown in Figure 9b is about 0.35 pu and, therefore, only 3.5 times higher than the rated current. The magnitude depends on the exact time of connection and the corresponding angle of the phase voltage. The maximum inrush current results from a time of connection when the magnitude of the phase voltage is in its minimum. Figure 9a shows that the phase voltages $u_{2N,t}$ and $u_{3N,t}$ have a magnitude of about 50% and, therefore, the resulting inrush current is lower than the maximum expected one.

Building on the findings of Figure 9, it can be stated that a black-start of the Microgrid shown in Figure 8, with connected transformers of the wind park and the local grid, would lead to a current demand much higher than the current capabilities of the BESS. However, it is possible to gradually increase the output voltage of the BESS while all transformers are connected to the Microgrid. By doing so, inrush currents, which result from switching actions that connect feeders with corresponding transformers, can be prevented. A measurement of such a gradual increase of the supplying voltage is shown in Figure 10.

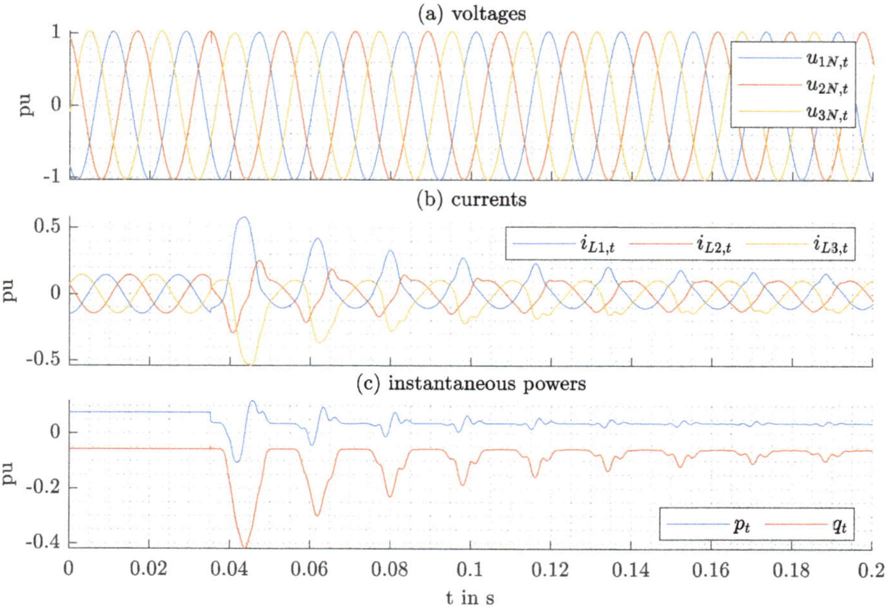

Figure 9. Measurement of the inrush current of the own auxiliary transformer of the Microgrid shown in Figure 8.

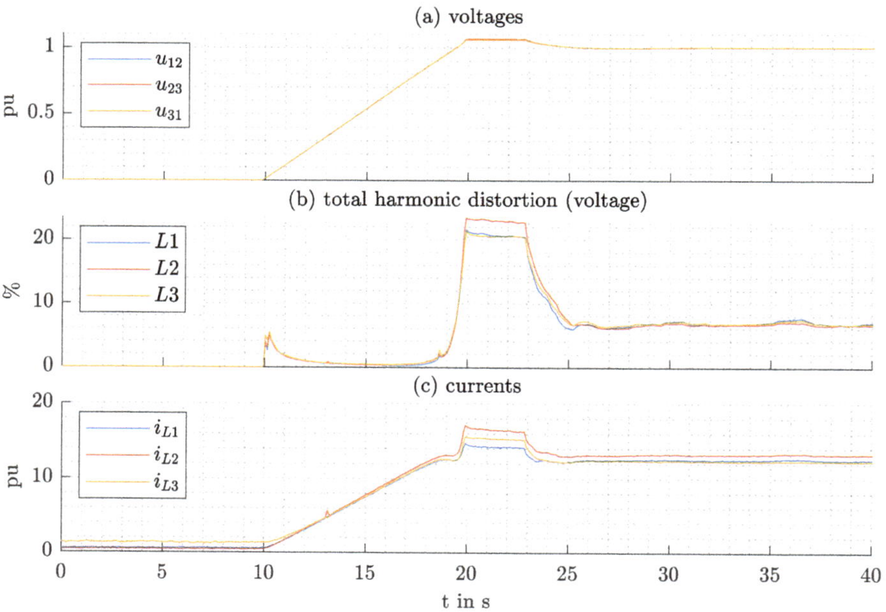

Figure 10. Measurement of voltages and current during a gradual increase of the voltage of the BESS in the Microgrid shown in Figure 8 while all transformers are connected. The small peak in i_{L2} can be considered as measurement error. Between 20 s–25 s, the voltage exceeds a value of 1 pu due to a high level of harmonics, indicated by the total harmonic distortion in (**b**), which leads to higher root-mean-square values of the voltages.

Figure 10 shows a gradual increase of the output voltage of the BESS within 10 s. Figure 10b shows that the current demand during this black-start of the Microgrid is only about 5% of the current capability of the BESS. About 15 s after the black-start, a stable supply voltage of 1 pu is reached and all feeders of the Microgrid are supplied.

3.2. Synchronization of the Wind Park to the Microgrid

After a black-start of the island grid, the wind turbines are able to synchronize with the supply voltage in the Microgrid. A wind turbine follows a start-up routine when it is activated from a state where it is switched off. The first step in this routine is the supply of its auxiliary systems, which, for example, includes the motors of the pitch control of the rotor blades (pitch drive) and the motor which aligns the wind turbine in the wind direction (azimuth drive). In the second step, the mechanical brake of the rotor of the wind turbine is released. In this "idle mode", the rotor rotates dependent on the wind speed, but no power infeed takes place. The measurement of a constant wind speed or a corresponding angular frequency of the rotor over a time period of several minutes, which lies above a minimum threshold, heralds the third step of the start-up routine. In the third step, the actual synchronization takes place, where the converter of the wind turbine synchronizes with the supply voltage in the Microgrid and starts to feed in active power, dependent on the actual wind speed and the corresponding angular frequency of the rotor.

3.3. Island Operation with an Active Frequency Control Characteristic

The frequency control characteristic shown in Figure 1 has been described as a simple possibility to integrate DER in the Microgrid based on a master–slave control approach. Such a frequency control characteristic was implemented in the BESS of the Microgrid by simply altering the value $\omega = 2\pi f(p, SoC)$ of the grid synchronization shown in Figure 2 according to the curve in Figure 1, which leads to a grid angle $\theta = 2\pi \int f(p, SoC)\, dt$. The behavior of such a control approach was measured during island operation after successfully synchronizing the wind park. The wind turbines in Microgrid support LFSMO with corresponding options for its parameterization. This parameterization was carried out according to the frequency curve shown in Figure 1 via several parameterizable fulcrums. The reaction of an activated LFSMO on the wind turbines works as a percentual power reduction in relation to the reference operating point of the wind turbine. For example, assuming a wind speed of $10\,\mathrm{m\,s^{-1}}$ leads to a reference operating point of the wind turbine of 50% of its nominal power, this reference operating point is executed directly as actual power output at 50 Hz and this reference operating point is executed as reduced power output point according to the frequency characteristic shown in Figure 1 at higher frequencies.

A measurement result of an islanded Microgrid operation, during which the frequency control approach mentioned above is active and the wind turbines are synchronized and feeding in power, is shown in Figure 11.

The field test started with a SoC of the BESS of 50% and an active power infeed of the wind park. Figure 11b shows the stepwise increase of the SoC of the BESS, which has been achieved by a manual narrowing of the usable SoC-window of the BESS. This manual change of the SoC was applied in order to shorten the time period of the field test. Due to the stepwise change of the SoC, the frequency needs some time to reach the corresponding reference value of the frequency control characteristic. The vertical lines represent these time periods. During these gradual changes of the frequency, the wind turbines reduce their currently possible power output P'_{DER} to a reduced power output P_{DER} according to the corresponding reference curve, which is shown in Figure 11a. During the time period of this gradual change of the frequency, the wind speed can be considered constant. In contrast, the time between manual changes of the usable SoC-window is quite long. Therefore, during these manual changes of the usable SoC-window, the wind speed and the corresponding power infeed of the wind park change. The horizontal lines of the power infeed in Figure 11a are related to these changes of the wind speed and the corresponding operating point P'_{DER} of the wind park. Nevertheless, Figure 11 proves to be a successful application of the frequency control approach according to Figure 1.

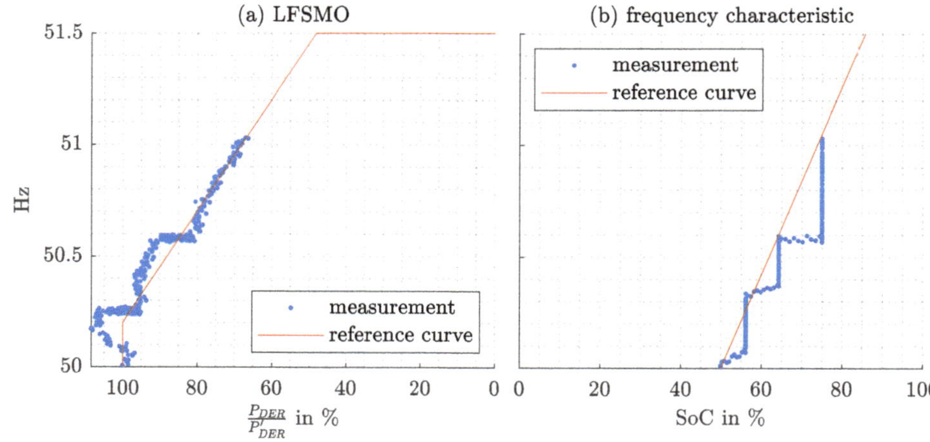

Figure 11. Measurement of the frequency control characteristic according to Figure 1.

3.4. Load Changes during Island Operation

The local grid of the Microgrid shown in Figure 8 is used to imitate significant load increases of a residential low voltage grid. The active power load bank consists of several resistors, which are connected to each other depending on which active power consumption is to be achieved. Figure 12 shows the measurement results of two load-steps of 200 kW and 800 kW and their influence on the voltage at the PCC of the BESS. During these load changes, the wind turbines were deactivated. Since the frequency characteristic in Figure 1 only influences the frequency when $p > 0 \wedge p > p_0$ or when $SoC > 0 \wedge SoC > SoC_0$ and a load-step leads to $p < 0$, the frequency during the load changes can therefore be considered to be constant $f = 50\,\mathrm{Hz}$ when assuming a state of charge of about $SoC = 50\%$.

The load-steps are coupled with a highly dynamic voltage drop, similar to the simulation results shown in Figure 5. However, the stabilization of the voltage to the nominal value takes about 4 s for both load-steps, which is much longer than observed in the simulations. Compared to the model used in the simulations, the control algorithms in the real converter are more complex and additional output filters are installed, which both explain the differences between field test and simulation. The voltage in Figure 5 does not fall below the value of of 0.8 pu, which is considered as the minimum dynamic voltage drop according to Bagert et al. [26].

The asynchronous machine of the local grid shown in Figure 8 is used to imitate massive reactive load changes of a typical low voltage grid. Figure 13 shows the measurement results of the start-up of two asynchronous machines with a nominal power of 50 kW and 200 kW and their influence on the voltage at the PCC of the BESS. The measurement results show a similar behavior compared to the simulation results shown in Figure 7.

As already described for the active power load changes the differences between the measurement- and simulations results can be explained by the much more complex control structure of the real converter.

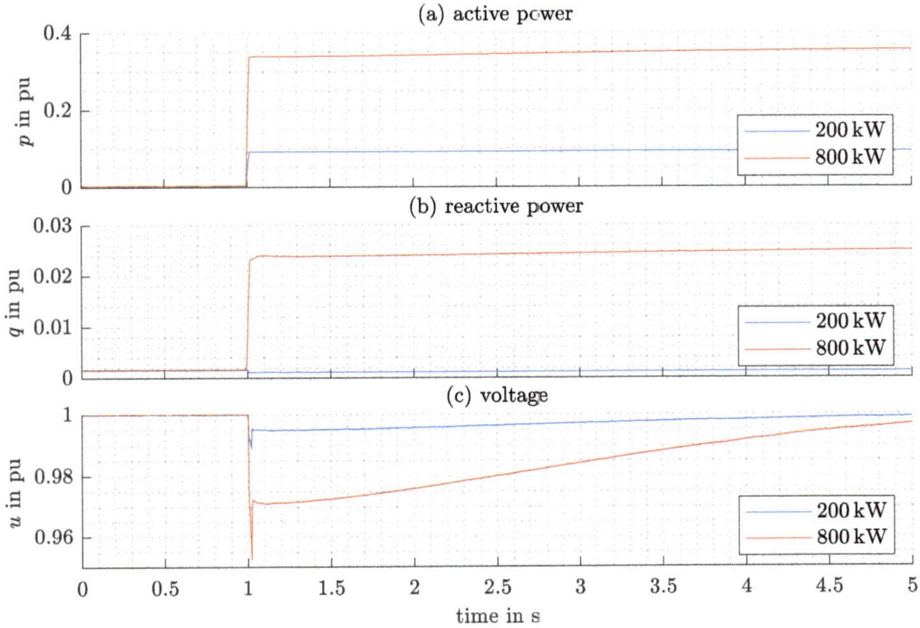

Figure 12. Measurements during an active power load-step of 200 kW and 800 kW. The voltage u is the mean value of all normalized line-to-line voltages.

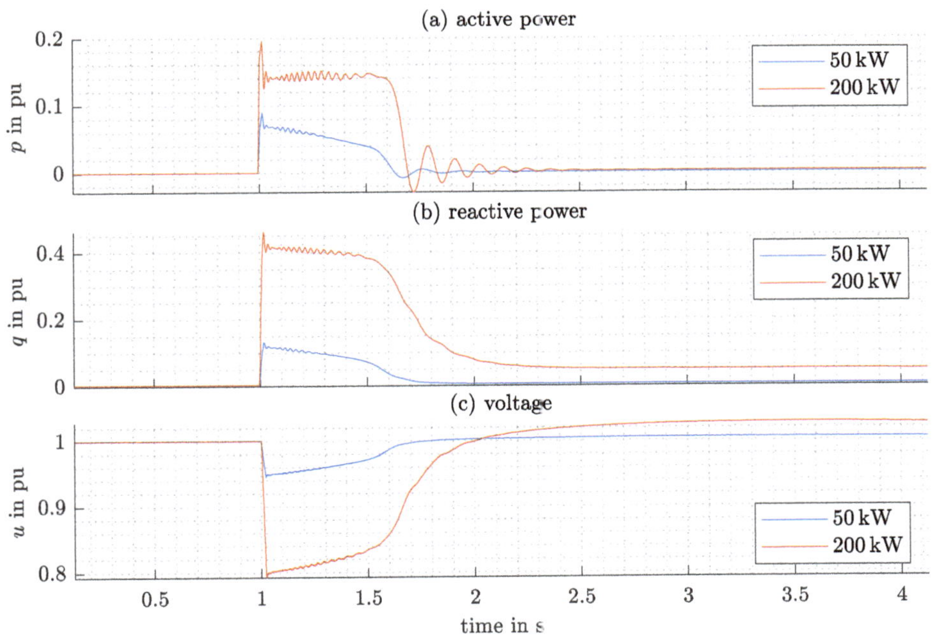

Figure 13. Measurements during the motor start-up with a nominal power of 50 kW and 200 kW.

3.5. Power Quality

Besides the voltage drops during load changes, which affect the power quality in a Microgrid, the occurrence of harmonics is another issue that affects the power quality. As shown in Figure 10, a high value of the total harmonic distortion (THD) may occur during the black-start of a Microgrid, which reaches a value of up to 20% in Figure 10b. Table 3 summarizes the THD for different operating points and situations during the field tests.

Table 3. Measured THD for different operating points and situations during the field tests.

Operating Point	Value of THD	Situation	Value of THD
$p = 0.08$ pu	2.5%	running 50 kW-motor	0.6%
$p = 0.16$ pu	2%	running 200 kW-motor	0.4%
$p = 0.24$ pu	1.5%	synchronization of wind turbines	15%
$p = 0.32$ pu	1%	$p = -0.2$ pu (moderate power infeed of wind turbines)	5%

The measurement results in Table 3 show a decreasing trend of the THD for an increasing value of the active power operating point p of the BESS. This can be explained by the fact that the associated converter of the BESS has been optimized for operation at $p = 0.8$ pu. Besides the rather high THD of 20% during the black-start of the Microgrid, as shown in Figure 10, during the synchronization of wind turbines, a similar high THD of 15% occurs. During the power infeed of wind turbines, the THD reaches a moderate level of 5%.

3.6. Discussion

In summary, the field tests prove that the BESS shown in Figure 8 is capable of performing a black-start of the Microgrid by using a gradual increase of the supply voltage during start-up. During island operation, the application of a frequency control characteristic based on a master–slave control approach allows the integration of DER with nominal powers higher than the BESS into the Microgrid, while also carrying out an appropriate energy management by guaranteeing the SoC of the BESS within allowable limits. The BESS is capable of controlling active- and reactive load changes, which are considered to be representative for significant load changes in a residential grid section, with an acceptable level of power quality, indicated by voltage drops not to fall below a value of 0.8 pu and by the THD, which only reaches high values during the black-start of the Microgrid and the synchronization of wind turbines.

4. Conclusions

In this paper, the black-starting and islanding capabilities of a BESS, which acts as grid-forming unit in a Microgrid, consisting of a wind park, a load and asynchronous machines, were investigated. The methodology for investigation is based on simulation in MATLAB/SIMULINK and the conduction of field tests. For the simulation, a simplified converter model is used, which allows the simulation of active- and reactive load changes during island operation. The simulation provides the basis for the selection of loads and motors for the field test and allows an estimation of the behavior during the field tests. Based on a master–slave control approach, a frequency characteristic is presented, which allows a proper power reduction of slave-units, without the further need for communication. The BESS represents the master-unit that controls the frequency, while DER represent slave-units, which vary their power output depending on the frequency. By applying this control approach, the operation and power infeed of DER with nominal powers higher than the BESS is possible, while also allowing an appropriate energy management to control the SoC of the BESS. An application of such an approach in Microgrids, which are "embedded" into the interconnected power grid, in order to ensure a backup supply during blackouts, therefore is possible very easily, as no further communication is necessary. By providing such an application of backup supply, a "value stacking" of BESS is possible and can increase the profitability of such a system when the service of backup supply is provided in addition to other services such as frequency control. In the field tests, the capabilities of the BESS to handle active- and reactive load changes were investigated. Furthermore, the master–slave control approach was implemented in the BESS and investigated during the field tests. After the successful black-start of the Microgrid, it could also be demonstrated that the BESS is capable of handling active- and reactive load changes and the master–slave control approach is applicable and works in a real Microgrid.

A future research goal is the investigation of protection schemes and their application in Microgrids supplying real low-voltage feeders instead of artificial ones.

Author Contributions: Conceptualization, J.M. and W.G.; Formal analysis, J.M.; Funding acquisition, W.G.; Methodology, J.M.; Software, J.M.; Supervision, W.G.; Visualization, J.M.; Writing—original draft, J.M.; Writing—review and editing, W.G. All authors have read and agreed to the published version of the manuscript.

Funding: This work received no external funding.

Acknowledgments: The publication of this article was supported by the Open Access Funding by TU Wien University Library.

Conflicts of Interest: The authors declare no conflict of interest.

Appendix A. List of Symbols

$\mathbf{i}_t$	normalized instantaneous current vector
$i_{L1,t}$	instantaneous current in the phase L1
$i_{L2,t}$	instantaneous current in the phase L2
$i_{L3,t}$	instantaneous current in the phase L3
$\underline{i}_S$	space vector of the current in the $\alpha\beta$-plane
$\underline{i}_{S,dq}$	space vector of the current in the dq-plane
$i_{S,d_{1+}}$	direct/active, positive-sequence component of the normalized output space vector of the current
$i_{S,q_{1+}}$	quadrature/reactive, positive-sequence component of the normalized output space vector of the current
$i_{S,d_{1+},ref}$	reference value of the direct/active, positive-sequence component of the normalized space vector of the current
$i_{S,q_{1+},ref}$	reference value of the quadrature/reactive, positive-sequence component of the normalized space vector of the current
$\mathbf{u}_{N,t}$	normalized phase-to-neutral instantaneous voltage vector
$\mathbf{u}_{N,t,ref}$	normalized phase-to-neutral instantaneous reference voltage vector
$u_{1N,t}$	normalized phase-to-neutral instantaneous voltage in the phase L1
$u_{2N,t}$	normalized phase-to-neutral instantaneous voltage in the phase L2
$u_{3N,t}$	normalized phase-to-neutral instantaneous voltage in the phase L3
$\underline{u}_{S,dq}$	normalized voltage space vector in the dq-plane
$\mathbf{U}$	root-mean-square value of the phase-to-phase voltage vector
U_n	nominal phase-to-phase voltage
$\mathbf{u}$	normalized phase-to-phase voltage vector
u_{12}	root-mean-square value of the normalized phase-to-phase voltage between the phases L1–L2
u_{23}	root-mean-square value of the normalized phase-to-phase voltage between the phases L2–L3
u_{31}	root-mean-square value of the normalized phase-to-phase voltage between the phases L1–L3
$u_{S,d_{1+}}$	direct/active, positive-sequence component of the normalized output space vector of the voltage
$u_{S,q_{1+}}$	quadrature/reactive, positive-sequence component of the normalized output space vector of the voltage
$u_{S,d_{1+},ref}$	reference value of the direct/active, positive-sequence component of the normalized output space vector of the voltage
$u_{S,q_{1+},ref}$	reference value of the quadrature/reactive, positive-sequence component of the normalized output space vector of the voltage
$\underline{u}_{1+}$	complex, normalized value of the positive-sequence voltage
u_{1+}	normalized magnitude of the positive-sequence voltage
θ	angle of the positive-sequence voltage
U_{DCL}	DC-link voltage
$u_{LC1N,t}$	instantaneous value of the phase-to-neutral voltage in the phase L1 at the inverter output
$u_{LC2N,t}$	instantaneous value of the phase-to-neutral voltage in the phase L2 at the inverter output
$u_{LC3N,t}$	instantaneous value of the phase-to-neutral voltage in the phase L3 at the inverter output
P_{DER}	power infeed of wind turbines

P_{LOAD}	power consumption of loads
P	electrical power of the converter
p	normalized active power output of the converter
q	normalized reactive power output of the converter
p_{ref}	reference value of the active power
q_{ref}	reference value of the reactive power
ω	angular frequency
ω_n	nominal angular frequency
L	inductance of the output filter of the converter
l	normalized inductance of the output filter of the converter
R	resistance of the output filter of the converter
C	capacitance of the output filter of the converter
c	normalized capacitance of the output filter of the converter
τ_i	time constant of the current control loop
$K_{I,P}$	proportional controller gain of the current control loop
$K_{I,I}$	integral gain of the current control loop
$K_{U,P}$	proportional gain of the voltage control loop
$K_{U,I}$	integral gain of the voltage control loop
S_n	nominal apparent power of the converter
Φ_R	phase margin of the converter control
p_t	instantaneous active power
q_t	instantaneous reactive power
u_C	normalized voltage at the PCC of the converter
u_{load}	normalized voltage at the connection point of the load
R'_{l1}	length-related resistance of line 1
X'_{l1}	length-related reactance of line 1
R'_{l2}	length-related resistance of line 2
X'_{l2}	length-related reactance of line 2
u_k	short circuit voltage of the converter transformer
P_k	short circuit power of the converter transformer
n	number of converters connected in parallel
f_{IL1}	switching signal for the switch S_{IL1}
f_{IL2}	switching signal for the switch S_{IL2}
f_{IL3}	switching signal for the switch S_{IL3}
f'_{IL1}	switching signal for the switch S'_{IL1}
f'_{IL2}	switching signal for the switch S'_{IL2}
f'_{IL3}	switching signal for the switch S'_{IL3}

References

1. Fraunhofer IWES. *The European Power System in 2030: Flexibility Challenges and Integration Benefits. An Analysis with a Focus on the Pentalateral Energy Forum Region*. Analysis on Behalf of Agora Energiewende. 2015. Available online: https://www.agora-energiewende.de/fileadmin2/Projekte/2014/Ein-flexibler-Strommarkt-2030/Agora_European_Flexibility_Challenges_Integration_Benefits_WEB_Rev1.pdf (accessed on 23 June 2020).
2. IRENA. Renewable Energy Prospects for the European Union. In *International Renewable Energy Agency (IRENA)*; European Commission (EC): Abu Dhabi, UAE, 2018. Available online: https://www.irena.org/publications/2018/Feb/Renewable-energy-prospects-for-the-EU (accessed on 23 June 2020).
3. Sterner, M.; Stadler, I. *Handbook of Energy Storage: Demand, Technologies, Integration*; Springer: Berlin/Heidelberg, Germany, 2019.
4. Marnay, C.; Chatzivasileiadis, S.; Abbey, C.; Iravani, R.; Joos, G.; Lombardi, P.; Mancarella, P.; von Appen, J. Microgrid evolution roadmap. In Proceedings of the 2015 International Symposium on Smart Electric Distribution Systems and Technologies (EDST), Vienna, Austria, 8–11 September 2015; pp. 139–144.
5. Mahmoud, M.S. *Microgrid: Advanced Control Methods and Renewable Energy System Integration*; Elsevier: Amsterdam, The Netherlands, 2016.
6. Tabatabaei, N.M.; Kabalci, E.; Bizon, N. *Microgrid Architectures, Control and Protection Methods*; Springer: Berlin/Heidelberg, Germany, 2019.

7. Roggatz, C.; Power, M.; Singh, N. Power System Restoration: Meeting the Challenge to Resiliency from Distributed Generation. *IEEE Power Energy Mag.* **2020**, *18*, 31–40. [CrossRef]
8. Zhao, B.; Zhang, X.; Li, P.; Wang, K.; Xue, M.; Wang, C. Optimal sizing, operating strategy and operational experience of a stand-alone microgrid on Dongfushan Island. *Appl. Energy* **2014**, *113*, 1656–1666. [CrossRef]
9. Katiraei, F.; Iravani, M.R. Power management strategies for a microgrid with multiple distributed generation units. *IEEE Trans. Power Syst.* **2006**, *21*, 1821–1831. [CrossRef]
10. Barklund, E.; Pogaku, N.; Prodanovic, M.; Hernandez-Aramburo, C.; Green, T.C. Energy management in autonomous microgrid using stability-constrained droop control of inverters. *IEEE Trans. Power Electron.* **2008**, *23*, 2346–2352. [CrossRef]
11. Peng, Z.; Wang, J.; Bi, D.; Wen, Y.; Dai, Y.; Yin, X.; Shen, Z.J. Droop control strategy incorporating coupling compensation and virtual impedance for microgrid application. *IEEE Trans. Energy Convers.* **2019**, *34*, 277–291. [CrossRef]
12. Han, H.; Liu, Y.; Sun, Y.; Su, M.; Guerrero, J.M. An improved droop control strategy for reactive power sharing in islanded microgrid. *IEEE Trans. Power Electron.* **2014**, *30*, 3133–3141. [CrossRef]
13. Garde, R.; Casado, S.; Santamaria, M.; Aguado, M. Power quality and stability analysis during islanded mode operation in a microgrid based on master–slave configuration. In Proceedings of the 2015 Saudi Arabia Smart Grid (SASG), Jeddah, Saudi Arabia, 7–9 December 2015; pp. 1–8.
14. Talapur, G.G.; Suryawanshi, H.M.; Shitole, A.B.; Nachankar, P. Combined Droop and Master-Slave Method for Load Sharing in Stand-alone AC Microgrid. In Proceedings of the IECON 2018-44th Annual Conference of the IEEE Industrial Electronics Society, Washington, DC, USA, 21–23 October 2018; pp. 1705–1710.
15. Verma, V.; Talpur, G.G. Decentralized master–slave operation of microgrid using current controlled distributed generation sources. In Proceedings of the 2012 IEEE International Conference on Power Electronics, Drives and Energy Systems (PEDES), Bengaluru, India, 16–19 December 2012; pp. 1–6.
16. Caldognetto, T.; Tenti, P. Microgrids operation based on master–slave cooperative control. *IEEE J. Emerg. Sel. Top. Power Electron.* **2014**, *2*, 1081–1088. [CrossRef]
17. Divakaran, A.M.; Hamilton, D.; Manjunatha, K.N.; Minakshi, M. Design, development and thermal analysis of reusable Li-ion battery module for future mobile and stationary applications. *Energies* **2020**, *13*, 1477. [CrossRef]
18. Friend, F. Cold load pickup issues. In Proceedings of the 2009 62nd Annual Conference for Protective Relay Engineers, Austin, TX, USA, 30 March–2 April 2009; pp. 176–187.
19. E-Control. Technische und organisatorische Regeln für Betreiber und Benutzer von Netzen: TOR Erzeuger: Anschluss und Parallelbetrieb von Stromerzeugungsanlagen des Typs B. 2019. Available online: https://www.e-control.at/documents/1785851/1811582/TOR+Erzeuger+Typ+B+V1.0.pdf/a9a7e5ae-5842-caa9-d2c0-93be4b6e0802?t=1562757801048 (accessed on 23 June 2020).
20. SMA. Design of Off-Grid Systems with Sunny Island 4.4M/6.0H/8.0H Devices. 2020. Available online: https://files.sma.de/downloads/Designing-OffGridSystem-PL-en-24.pdf (accessed on 23 June 2020).
21. Marchgraber, J.; Gawlik, W. Dynamic Voltage Support of Converters during Grid Faults in Accordance with National Grid Code Requirements. *Energies* **2020**, *13*, 2484. [CrossRef]
22. Marchgraber, J.; Alács, C.; Guo, Y.; Gawlik, W.; Anta, A.; Stimmer, A.; Lenz, M.; Froschauer, M.; Leonhardt, M. Comparison of Control Strategies to Realize Synthetic Inertia in Converters. *Energies* **2020**, *13*, 3491. [CrossRef]
23. Henninger, S. Netzdienliche Integration Regenerativer Energiequellen über Stromrichtergekoppelte Einspeisenetze mit Integrierten Energiespeichern. Ph.D. Thesis, FAU Erlangen, Erlangen, Germany, 2019.
24. Yazdani, A.; Iravani, R. *Voltage-Sourced Converters in Power Systems: Modeling, Control, and Applications*; John Wiley & Sons: Hoboken, NJ, USA, 2010.
25. Wurm, M.; Jonke, P.; Marchgraber, J.; Gawlik, W.; Vitovec, W. Ortsnetz-Inselbetriebsversuch mit einem 2,5-MVA/2,2-MWh-Batteriespeicher: Messergebnisse und Vergleich mit einem Controller Hardware-in-the-loop Setup. *E I Elektrotechnik Und Informationstechnik* **2019**, *136*, 368–376. [CrossRef]

26. Bagert, M. *Elektrischer Eigenbedarf: Elektrotechnik in Kraftwerken und Industrie, Zusammenwirken von Kraftwerken und Netzen, Beispiele Ausgeführter Anlagen*; VDE Verlag: Berlin, Germany, 2012.
27. Marchgraber, J.; Alács, C.; Nemec-Begluk, S.; Gawlik, W.; Jonke, P.; Wurm, M.; Wailzer, G.; Vitovec, W. Schwarzstart und Inselbetrieb eines Netzabschnitts mit Windenergieeinspeisung mithilfe eines Batteriespeichers. In Proceedings of the 11th Internationale Energiewirtschaftstagung, Vienna, Austria, 13–15 February 2019.

energies

MDPI

Article

Real-Time Validation of Power Flow Control Method for Enhanced Operation of Microgrids

Hossein Abedini [1,2,*], Tommaso Caldognetto [1,2], Paolo Mattavelli [1,2,*] and Paolo Tenti [2,3]

1 Department of Management and Engineering, University of Padova, 36100 Vicenza, Italy; tommaso.caldognetto@unipd.it
2 Interdepartmental Centre Giorgio Levi Cases, University of Padova, 35131 Padova, Italy; paolo.tenti@unipd.it
3 Department of Information Engineering, University of Padova, 35131 Padova, Italy
* Correspondence: hossein.abedini@unipd.it (H.A.); paolo.mattavelli@unipd.it (P.M.)

Received: 15 October 2020; Accepted: 12 November 2020; Published: 15 November 2020

Abstract: This paper describes a control methodology for electronic power converters distributed in low-voltage microgrids and its implementation criteria in general microgrid structures. In addition, a real-time simulation setup is devised, implemented, and discussed to validate the control operation in a benchmark network. Considering these key aspects, it is shown that operational constraints regarding the power delivered by sources, flowing through network branches, and exchanged at the point of connection with the main grid can generally be fulfilled by the presented control approach. The control is performed considering a cost function aiming at optimizing various operation indexes, including distribution losses, current stresses on feeders, voltage deviations. The control system allows an enhanced operation of the microgrid, specifically, it allows dynamic and accurate power flow control enabling the provision of ancillary services to the upstream grid, like the demand–response, by exploiting the available infrastructure and the energy resources. Then, the validation of the approach is reported by using a real-time simulation setup with accurate models of the power electronic converters and related local controllers, of the grid infrastructure, of the power flow controller, and of the communication network used for data exchange. It is also shown that the implemented platform allows to fully reproduce, analyze, and finally validate all the relevant steady-state and dynamic behaviors related in the considered scenario.

Keywords: demand–response; distributed electronic power converters; optimal power sharing; power flow control; real-time simulations

1. Introduction

The role of distribution networks in power system management and support is changing dramatically. Revisions of the market framework are expected in the near future in order to exploit distributed resources for supporting upstream medium and high voltage grids [1,2]. In the perspective envisioned by the European directive [3], microgrids will be the bricks of future electric systems. They embrace loads and sources that are close to each other and can be synergized to pursue a safe and cost-effective operation of the electric system and the innovative feature of allowing end-users to become actors of the electricity market. For this aim, microgrids are expected to evolve into systems capable to ensure degrees of scalability, flexibility, reliability, robustness, and readiness similar to networks of digital devices. From the perspective of this analogy, we may refer to E-LAN, namely, Local Area Energy Networks [4]. E-LANs, represented in Figure 1, allow important features, including optimal power flow control, dispatchability at multiple points of connections with the main grid, exploitation of all the energy resources available. In general, advanced control features rely on adequate information and telecommunication (ICT) infrastructures [5]. Such infrastructures are an important constituent of modern intelligent energy systems, whose impact, also in case of malfunctions, is rarely included in studies considering low-voltage distribution grids. Indeed, from the perspective of the required communication protocols and specifications, the considered scenario shows still fluid and evolving [6].

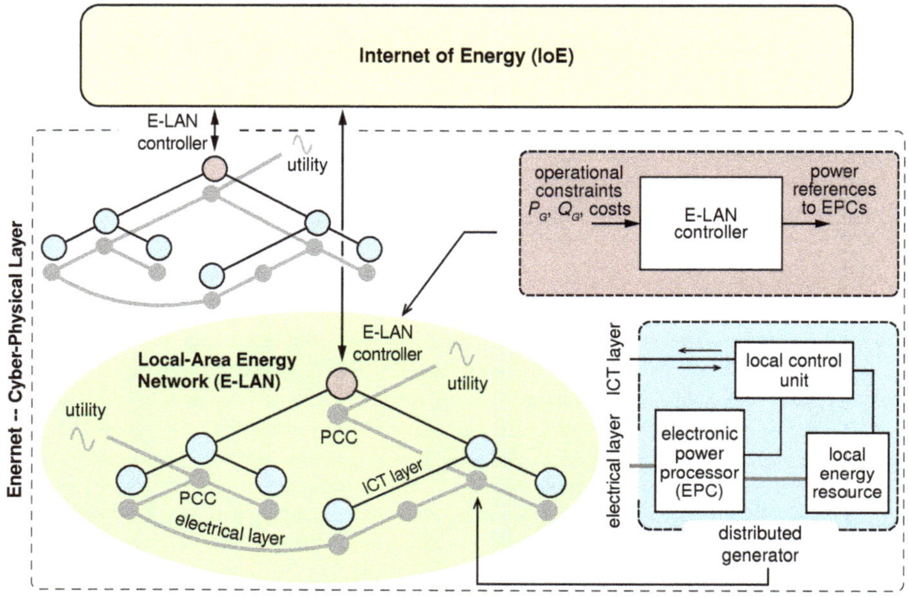

Figure 1. Modern power system scenario.

Various contributions align along the outlined direction organize microgrid control in multiple layers, as shown in Figure 2. The planning of the resources based on energy arbitrage is found at the higher level of the E-LAN control hierarchy, that is, the transactive control layer, which can be applied both at the microgrid level [1,7,8], also by exploiting detailed mathematical modeling of the distributed energy resources [9], and at the premises of single consumers too [10]. By these approaches, predictions about power needs and energy prices are taken into account to optimally exploit power flow control, as specifically done in [1,8,9]. On the other hand, network models and power flow constraints, which are considered herein, are crucial for optimal utilization of the microgrid distribution infrastructure [11]. This is particularly important when ancillary services, like demand–response, involving additional constraints to be met, have to be accommodated by relying on distributed energy resources interfaced by electronic power converters (EPCs). From this respect, automatic and predetermined power sharing techniques, see, for example, [12], typically constituting the primary control layer of microgrids [13,14], should be augmented to adapt to actual power needs and fulfill given power flow constraints optimally. A contribution from this perspective is given in [4], in which an optimal power flow controller is proposed considering steady-state operation. Herein, the approach is revised and implemented on a real-time simulation platform to evaluate its operation in dynamic conditions. Of course, such approaches may be applied jointly with load prioritization techniques based on load analyses, as proposed in [15]. These techniques can schedule the on/off status of the loads to be supplied by the available sources. The available sources can then be coordinated by optimal power flow control signals. The optimization approach described herein aims at taking advantage of every source available in the grid without using power shedding methods except those enforced at higher levels of the control hierarchy. Instead, enhanced performance of the network is pursued by synergistic use of the control abilities of any distributed power sources.

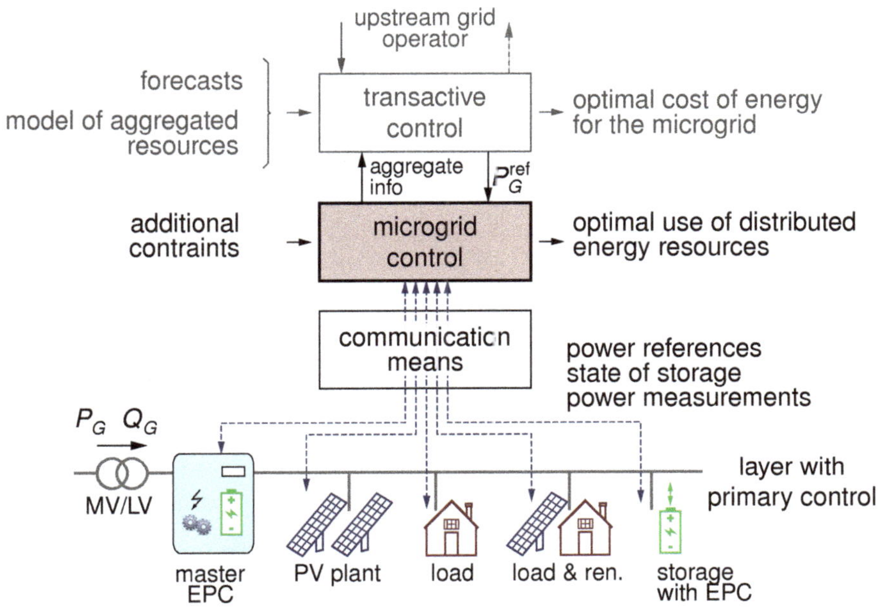

Figure 2. Local Area Energy Network (E-LAN) control structure.

The validation of the approaches mentioned above is a delicate task due to the complexity and the variety of the dynamics involved (e.g., the fast response of EPCs versus the slow optimization processes and power system dynamics). For these reasons, especially when particularly complex systems are analyzed, validation is often performed by means of computer simulations rather than experimental prototype realizations. Thanks to the recent advances in digital computing, real-time simulators have been employed lately for systems studies involving the interaction of power systems and power electronics systems that are characterized by fast dynamics (e.g., tens of μs) [16]. Validation via real-time simulations presents several advantages as compared to traditional simulations. The principal ones are (*i*) the possibility of performing an on-line testing of models and controls, even while interacting with other hardware components or prototypes [17], and (*ii*) the possibility of emulating parts of a complex experimental scenario that may not be conveniently included otherwise, due to size, cost, safety, or availability constraints. Several hardware solutions are available to run real-time simulations. Some exploit general purpose toolsets, as shown in [18] and, before, in [19] and [20], others use dedicated hardware and software solutions to ease the development of models of electrical and electronic systems, as done, for example, in [21–23].

In this paper, a control architecture that makes use of an innovative optimization framework capable to fulfill the operating constraints while providing synergistic operation of all controllable sources acting in the grid is considered and analyzed. The system performances are optimized in terms of component stress, power sharing, voltage stability, energy efficiency, congestion management, demand–response, robustness against transients, and communication failures. The proposed control is tested by a real-time simulation setup combining real-time simulators (OPAL-RT), industrial central controllers, and communication network emulators. These two aspects constitute the contributions of the paper, that is, (*i*) describe a power flow optimization method applied dynamically to fulfill power constraints, and (*ii*) describe the implementation of the proposed power flow control considering a real-time simulation setup integrating fully modeled converters controllers, realistic communication performance, allowing to validate the effectiveness of the proposed optimal control.

In the reminder of the paper, the power flow control is presented in Section 2, while Section 3 presents the primary-local control of the distributed electronic power converters. The implementation of the whole control system is described in Section 4, which also reports and discusses the obtained results. Section 5 concludes the paper.

2. Coordination of Distributed Electronic Power Converters

The power flow control method considered herein is introduced in the following. The method allows to satisfy various operational constraints by exploiting the available distributed EPCs in an optimal way. The method is validated for the first time in this paper by means of real-time simulations and shown suitable for real-time control. The results are reported in Section 4.

2.1. Network Equations

Consider an electrical grid, either single-phase or three-phase, with L branches and N nodes, plus the slack node (node 0) whose voltages v_0 are taken as reference voltages. Loads and sources are connected phase-to-phase or phase-to-neutral, while network branches interconnect pairs of nodes. The network graph is described by the $L \times N$ incidence matrix $\mathbf{A}$, where the column corresponding to node 0 is omitted. For simplicity, but without loss of generality, in the following we will refer to a single-phase network where loads and sources are connected between the grid nodes and a common ground, corresponding to the neutral wire.

Let $\boldsymbol{u}$ be the (vector of) node voltage deviations from reference v_0, $\boldsymbol{i}$ the currents entering the grid nodes, $\boldsymbol{w}$ the voltages across the branches oriented according to the network graph, and $\boldsymbol{j}$ the corresponding branch currents. The Kirchhoff's laws give:

$$\boldsymbol{w} = \mathbf{A}\boldsymbol{u}, \quad \boldsymbol{i} = \mathbf{A}^T\boldsymbol{j}, \tag{1}$$

where superscript T denotes transposition. In sinusoidal operation, we represent currents and voltages as phasors and correspondingly we may define the diagonal matrix $\boldsymbol{\Lambda}$ of branch impedances. Correspondingly, the relations between branch currents and voltages become:

$$\boldsymbol{w} = \boldsymbol{\Lambda}\boldsymbol{j}, \quad \boldsymbol{j} = \boldsymbol{\Lambda}^{-1}\boldsymbol{w}. \tag{2}$$

The relations between node voltages and currents are the following:

$$\boldsymbol{i} = \mathbf{Y}\boldsymbol{u}, \quad \boldsymbol{u} = \mathbf{Y}^{-1}\boldsymbol{i} = \mathbf{Z}\boldsymbol{i}, \quad \text{where:} \qquad \mathbf{Y} = \mathbf{A}^T\boldsymbol{\Lambda}^{-1}\mathbf{A}. \tag{3}$$

In (3), $\mathbf{Y}$ is the nodal admittance matrix and its inverse $\mathbf{Z}$ is the nodal impedance matrix. Finally, we get the inverse relations of (1) by:

$$\boldsymbol{u} = \mathbf{B}\boldsymbol{w} \quad \boldsymbol{j} = \mathbf{B}^T\boldsymbol{i} \quad \text{where} \quad \mathbf{B} = \mathbf{Z}\mathbf{A}^T\boldsymbol{\Lambda}^{-1}. \tag{4}$$

Remarkably, such equations apply to both meshed and radial networks. In this latter case, $\mathbf{B} = \mathbf{A}^{-1}$.

2.2. Control Equations

Figure 3 schematically represent a network, indicating the kind of nodes and related referred to thereafter. In general terms, the nodes can be classified as:

(a) *Voltage nodes*, supplied by voltage sources. Let $\boldsymbol{u}_v$ be the voltages imposed at these nodes, referred to the reference voltage v_0. Currents $\boldsymbol{i}_{vs}$ supplied by the sources partially feed local loads ($\boldsymbol{i}_{vl}$) and partially enter the grid ($\boldsymbol{i}_v$).

(b) *Current nodes*, supplied by current sources. Currents $\boldsymbol{i}_{cs}$ supplied by the sources partially feed local loads ($\boldsymbol{i}_{c\ell}$) and partially enter the grid ($\boldsymbol{i}_c$). Let $\boldsymbol{u}_c$ be the voltages at such nodes.

(c) *User nodes*, supplying passive loads. Let u_u be the voltages at such nodes and $i_{u\ell}$ be the related load currents.

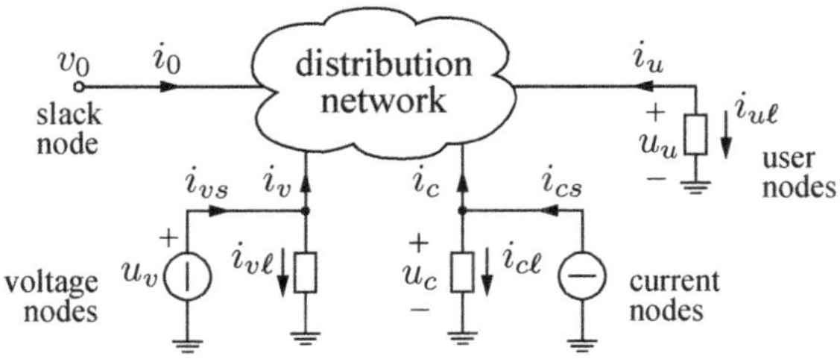

Figure 3. General network representation.

The function of balancing the local generation by sources and consumption by possibly connected local loads is in charge of the local controller of the current or voltage nodes. The local controller, depending on the requests coming from the central controller, may adapt its power generation to fully compensate for local load or to pursue local optimization criteria.

It is easy to show that all network voltages and currents can be expressed as a function of voltages u_v and currents i_c impressed by the sources which, in turn, can be controlled by acting on the EPCs interfacing the sources with the grid.

In the following, we therefore consider the voltages u_v impressed at voltage nodes, and the currents i_c entering the grid at current nodes, as the control (input) variables for the entire grid. The main output variables are currents i_v at voltage nodes and voltages u_c at current nodes, all remaining grid quantities being easily derived.

We generally express the control-to-output equations in the form:

$$\begin{bmatrix} i_v \\ u_c \end{bmatrix} = \mathbf{H} \begin{bmatrix} u_v \\ i_c \end{bmatrix} = \begin{bmatrix} \mathbf{H}_{vv} & \mathbf{H}_{vc} \\ \mathbf{H}_{cv} & \mathbf{H}_{cc} \end{bmatrix} \begin{bmatrix} u_v \\ i_c \end{bmatrix} = \begin{bmatrix} \mathbf{Y}_{vv} - \mathbf{Y}_{vc}\mathbf{Y}_{cc}^{-1}\mathbf{Y}_{cv} & \mathbf{Y}_{vc}\mathbf{Y}_{cc}^{-1} \\ -\mathbf{Y}_{cc}^{-1}\mathbf{Y}_{cv} & \mathbf{Y}_{cc}^{-1} \end{bmatrix} \begin{bmatrix} u_v \\ i_c \end{bmatrix}, \tag{5}$$

where $\mathbf{Y}_{vv}$, $\mathbf{Y}_{vc}$, $\mathbf{Y}_{cv}$, and $\mathbf{Y}_{cc}$ are sub-matrices of $\mathbf{Y}$ in (3) that refer to voltage and current nodes, respectively, and $\mathbf{H}$ is the control-to-output transfer matrix.

From the above equations, we express the currents at voltage nodes as:

$$\begin{gathered} \mathbf{i}_v = \mathbf{H}_{vv}\mathbf{u}_v + \mathbf{H}_{vc}\mathbf{i}_c + \mathbf{i}_v^0 \quad \text{where:} \\ \mathbf{H}_{vv} = \mathbf{Z}_{vv}^{-1}, \qquad \mathbf{H}_{vc} = -\mathbf{Z}_{vv}^{-1}\mathbf{Z}_{vc}, \qquad \mathbf{i}_v^0 = -\mathbf{Z}_{vv}^{-1}\mathbf{Z}_{vu}\mathbf{i}_{u\ell} \end{gathered} \tag{6}$$

2.3. Constraints

In general, the grid control problem is twofold. On one side, we wish to optimize some aspects of grid operation, as explained in the following section. On the other side, we need to fulfill specific constraints in terms of power flow at a given set of grid nodes or branches.

More specifically, in order to control the active and reactive power entering the grid at voltage nodes, currents i_{vs} can be constrained. In particular, constraints may apply to their direct (active) and/or quadrature (reactive) terms. Let:

$$i_{vs} = i_{vsd} + i_{vsq} \tag{7}$$

we assume that, among the N_v currents $\boldsymbol{i}_{vs}$ fed by voltage sources, N_δ are subject to constraints on the direct component, and N_γ are subject to constraints on the quadrature component. Let $\boldsymbol{i}_{v\delta}$ and $\boldsymbol{i}_{v\gamma}$ be such constrained currents, the constraints are expressed as (superscript ref indicates reference values):

$$\begin{cases} \boldsymbol{\Psi}_{v\delta} = \boldsymbol{i}_{v\delta} - \boldsymbol{i}_{v\delta}^{\text{ref}} = \boldsymbol{0}_{N_{v\delta}} \\ \boldsymbol{\Psi}_{v\gamma} = \boldsymbol{i}_{v\gamma} - \boldsymbol{i}_{v\gamma}^{\text{ref}} = \boldsymbol{0}_{N_{v\gamma}} \end{cases} \tag{8}$$

Similar constraints can also apply to the direct and quadrature currents entering the grid at slack node, which are related to the active and reactive power P_G^{ref} and Q_G^{ref} at the point of coupling with the upstream grid.

Currents $\boldsymbol{i}_{cs}$ fed by current sources can also be subject to constraints, expressed by:

$$\begin{cases} \boldsymbol{\Psi}_{c\delta} = \boldsymbol{i}_{c\delta} - \boldsymbol{i}_{c\delta}^{\text{ref}} = \boldsymbol{0}_{N_{c\delta}} \\ \boldsymbol{\Psi}_{c\gamma} = \boldsymbol{i}_{c\gamma} - \boldsymbol{i}_{c\gamma}^{\text{ref}} = \boldsymbol{0}_{N_{c\gamma}} \end{cases} \tag{9}$$

In practice, constraints in (9) reduce the number of control variables, freezing a subset of impressed currents $\boldsymbol{i}_\delta$ and $\boldsymbol{i}_\gamma$.

A last type of constraint may impose specific values to a set of branch currents. This corresponds to enforce the power flow in specific grid lines (power steering) or clearing specific branch currents (active insulation). Let N_j be the number of constrained branches, we may express these constraints, separately on d and q axes, as:

$$\begin{cases} \boldsymbol{\Psi}_{j\delta} = \boldsymbol{0}_{N_j} \\ \boldsymbol{\Psi}_{j\gamma} = \boldsymbol{0}_{N_j} \end{cases} \tag{10}$$

2.4. Cost Function

As mentioned before, the E-LAN control variables can be determined according to an optimal control approach, where a suitable cost function φ is minimized while fulfilling the above set of constraints.

In general terms, we define the cost function as:

$$\varphi = c_g \varphi_{grid} + c_c \varphi_{conv} + c_u \varphi_u \tag{11}$$

where coefficients c_g, c_c, c_u are weighting factors, and variables φ_{grid}, φ_{conv}, φ_u are the cost function terms, defined as follows.

- φ_{grid} corresponds to the power loss in the distribution grid, expressed in relative terms as:

$$\varphi_{grid} = \frac{P_{grid}}{P_{loss}^0} = \frac{\boldsymbol{r}^T \boldsymbol{J}^2}{P_g^0 + P_c^0} \tag{12}$$

  where P_{grid} is the grid loss in a generic operating condition, P_{loss}^0 is total power loss in the condition when all controllable quantities are set to zero, $\boldsymbol{J}^2$ is the vector of square rms values of branch currents, and $\boldsymbol{r}$ is the vector of branch resistances. P_{loss}^0 results by adding P_{grid}^0 (grid loss) and P_{conv}^0 (conversion loss).
- φ_{conv} corresponds to the total power loss in the EPCs interfacing the distributed generators with the grid, which can be driven as voltage sources or current sources to implement sources $\boldsymbol{u}_v$ and $\boldsymbol{i}_c$, respectively. It is expressed by:

$$\varphi_{conv} = \frac{P_{conv}}{P_{loss}^0} = \frac{\boldsymbol{r}_g^T \boldsymbol{I}_{gs}^2 + \boldsymbol{r}_s^T \boldsymbol{I}_s^2}{P_{grid}^0 + P_{conv}^0} \tag{13}$$

where P_{conv} is the conversion loss in a generic operating condition, $\boldsymbol{r}_g$ and $\boldsymbol{r}_s$ are the vectors of equivalent series resistances of voltage and current sources, $\boldsymbol{I}_{gs}^2$ and $\boldsymbol{I}_s^2$ are vectors of square rms values of source currents.

- φ_u corresponds to the cumulative rms deviation of node voltages from voltage reference v_0; it is given by the ratio between the square cumulative rms voltage deviation in a generic condition and the corresponding value when all controllable variables are set to zero:

$$\varphi_u = \frac{||\boldsymbol{u}_u||^2}{||\boldsymbol{u}_u^0||^2} = \frac{\sum_{n=1}^{N} U_{un}^2}{\sum_{n=1}^{N} {U_{un}^0}^2} \tag{14}$$

It is worth remarking that the coefficients in (11) may be tuned independently in order to assign different weights to voltage deviations, grid losses, and conversion losses on the optimization on the basis of the specific requirements of the application scenario.

In a similar way, we may extend the cost function to include other terms related to the power stress of distributed sources, the thermal stress of feeders. the VA stress of EPCs. The result is a cost function that accounts for the main operation aspects influencing the grid performance, and prevents useless stress of the grid components.

2.5. Solution of the Optimal Control Problem

Eventually, the grid control problem can be formulated as a constrained optimum problem, where cost function φ is minimized while fulfilling constraints $\boldsymbol{\psi}$:

$$\min \varphi(\boldsymbol{x}) \quad \text{such that} \quad \boldsymbol{\psi}(\boldsymbol{x}) = \boldsymbol{0} \tag{15}$$

where $\boldsymbol{\psi}$ represents the set of constraints expressed in the linear form:

$$\boldsymbol{\psi}(\boldsymbol{x}) = \mathbf{D}_{\psi}\boldsymbol{x} + \mathbf{E}_{\psi} \tag{16}$$

and the cost function is expressed in the quadratic form:

$$\varphi(\boldsymbol{x}) = \frac{1}{2}\boldsymbol{x}^T\mathbf{D}_{\varphi}\boldsymbol{x} + \boldsymbol{x}^T\mathbf{E}_{\varphi} + \varphi_0 \tag{17}$$

Expressing the matrices shown in (16) and (17) as a function of network quantities, the optimum control problem can solved in explicit form.

It can be observed that the above constraints do not include inequalities. Actually, the quantities that could be constrained by inequalities (e.g., current and power stresses, voltage deviations) are included in the cost function with proper weighting coefficients. The advantage of this approach is that the solution is found in explicit form, thus preventing convergence problems of the solving algorithm and making this latter very fast.

Finally, it is worth remarking that the above approach requires the knowledge of the network topology and network parameters. Actually, even if these data may be not fully available, there are methods presented in literature that allow identification of such information by measurements at grid nodes [24–26].

3. Local Control of Electronic Converters

The literature categorizes the behavior of distributed EPCs when taken singularly in grid-feeding, grid-supporting, and grid forming [27]. In the presence of a centralized microgrid controller dispatching optimal power commands to distributed EPCs, the grid-feeding behavior may be convenient, because it allows to easily operate multiple parallel-connected EPCs following given power references, regardless of grid parameters values. Instead, in case of accidental transition to islanded operation, the grid-supporting behavior may be the most favorable, because it allows sustaining the voltage of the local sub-grid section became isolated. This is often done by means of hierarchical structures, as described, for example, in [28], which, typically, aim at defining the operating voltage and frequency of the microgrid, but without specific load sharing schemes based on the available resources and network structure. The control structure in Figure 4, firstly presented in [29], combines valuable merits of grid-following and grid-supporting: it achieves output power regulation when the grid voltage if stiff and supports the grid voltage during transients and in case of transitions to the islanded operation.

Specifically, Figure 5 shows the complete structure of an EPC equipped with inner current and voltage controllers and P-f and Q-V droop loops. On top of these standard control loops, a power regulator with constrained output is employed. The two power control loops, that is, for the active and reactive powers, modify the droop characteristics by vertical shifts in order to make the converter follow given power references. In case of abnormal grid conditions that occur, for example, if the grid becomes particularly weak or islanded, the power controllers tend to saturate automatically. In this condition, the EPCs behave as traditional droop controlled converters, sharing incremental power needs in inverse proportion to their droop coefficients [27].

Remarkably, the considered control attains output power flow control while operating connected to the main grid, and autonomous operation with load sharing in case islanded operation occurs.

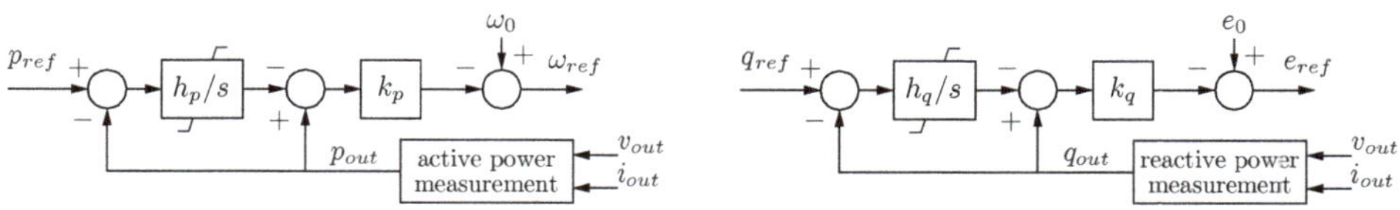

Figure 4. Droop control scheme with additional power control loop.

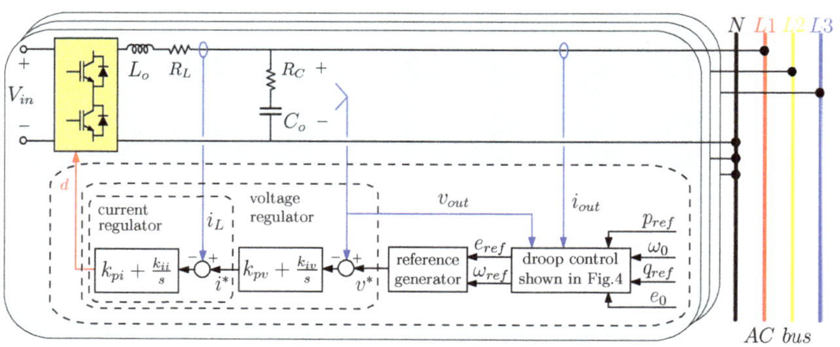

Figure 5. Control loops of distributed EPCs. comprising an inner inductor current control loop, voltage control loop, and droop laws.

4. Real-Time Simulation Results

The real-time simulation setup shown in Figure 6 has been implemented in order to evaluate the proposed approaches in steady-state and dynamic conditions. To this end, the benchmark low-voltage network proposed in [30] and arranged as indicated in Figure 7 is considered. Figure 8 displays its model on the real-time simulator.

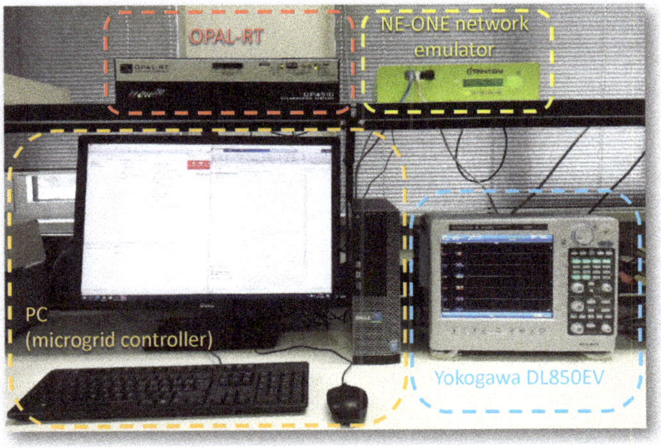

Figure 6. Real-time experimental setup with highlighted the OP4510 real-time simulator executing the network model and the EPCs control algorithms, the NE-ONE communication network emulator, the PC executing the microgrid controller, and the Yokogawa DL850EV for long-term data acquisition, processing, visualization, and recording.

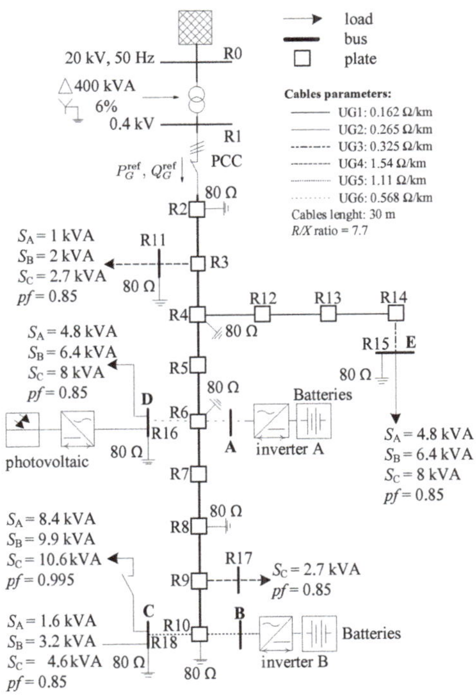

Figure 7. Considered low-voltage microgrid.

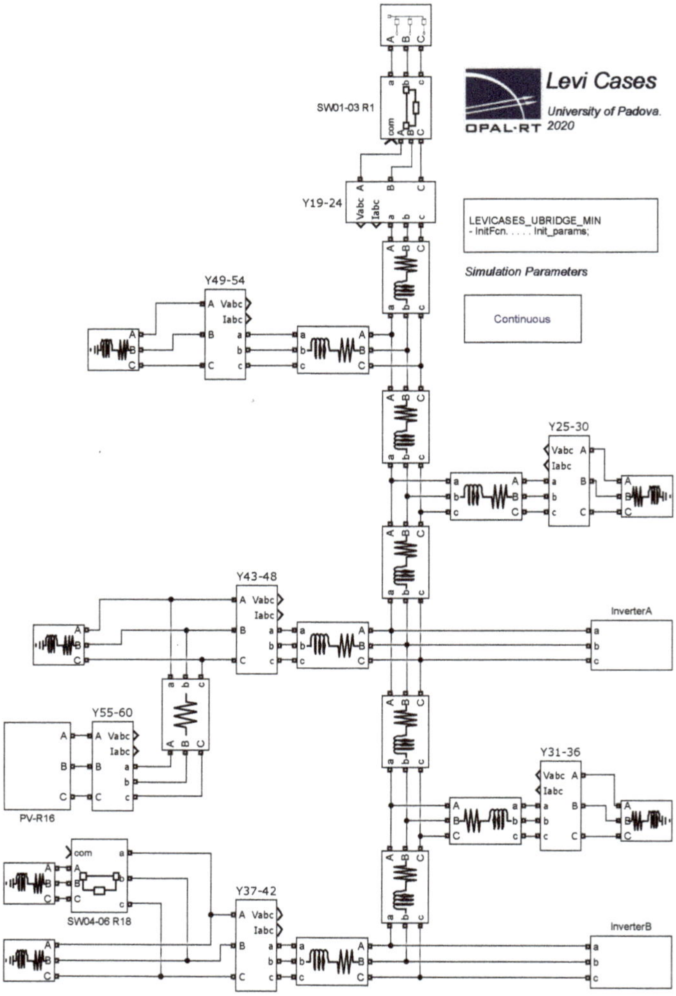

Figure 8. Network model executed in FPGA.

The setup is composed of an OPAL OP4510 real-time simulator to execute the network model and the EPCs control algorithms, an iTrinegy NE-ONE network emulator to emulate the features of a real communication network, a computer to execute the microgrid controller, and the Yokogawa DL850EV for long-term data acquisition, processing, visualization, and recording. The network and the EPCs hardware are implemented using the eHS Gen 4 Solver and executed, with a time-step of 2.5 µs, on the Xilinx FPGA board Kintex-7 325T embedded in the OP4510. EPC controls and communication interface are implemented using Simulink blocks and they executed, with a time-step of 50 µs, on the four cores of the 3.5-GHz Intel Xeon CPU embedded in the OP4510. Such a model partitioning allows performing accurate real-time simulations of the considered network and of the EPC's hardware on FPGA and, concurrently, execute more complex converters controllers on CPU. The microgrid controller for optimal power flow control is implemented in Matlab and executed, with execution frequency of 2 Hz, on the desktop computer. The performances of the used computer are reported by means of the vector returned by the Matlab command `bench`: [0.12 0.11 0.02 0.13 0.36 0.41]. Control settings and monitoring is allowed by a dedicated graphical user interface, displayed in Figure 9. The microgrid model that is emulated by the OP4510 and the power flow controller that is executed on the computer exchange information representing power-terms by UDP communication via the network emulator NE-ONE. The network emulator can reproduce ideal or impaired communication conditions by including delays, packet loss or corruption, latencies, etc., which allows validation with realistic communication network performances. Finally, data acquisition, visualization, and storage are performed by the DL850EV. This latter is also exploited for active and reactive power measurements using dedicated, embedded processing functions and real-time computation capabilities.

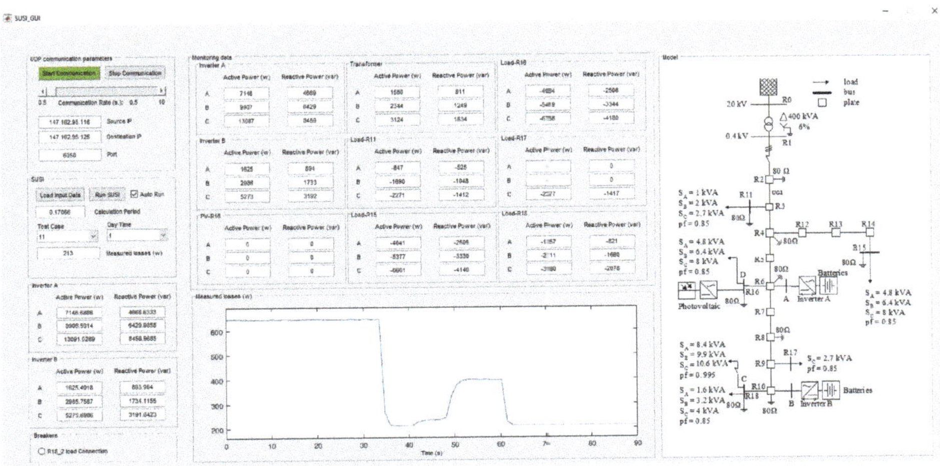

Figure 9. Graphical user interface for monitoring and control settings.

4.1. Model Structure Details

Figure 10 displays the whole real-time simulation model. The model is partitioned in two parts, one running on the FPGA, the other on the CPU. The FPGA partition contains the model of all the hardware parts, as shown in Figure 8, namely, the grid network specified in Figure 7 and the output filters of the EPCs hardware specified in Table 1. The CPU partition comprises the following blocks.

- Central_ctr_interaction is responsible for UDP ethernet communication between OP4510 and the microgrid controller running on the computer. The packet rate and control algorithm in execution are 500 ms; if needed, this time can be reduced to 10 ms with computers of higher performance or by dedicated, optimized implementations. This block also (*i*) logs the results from the running experiment, (*ii*) acquires and dispatches the control signals that are sent by the microgrid controller to the EPCs, (*iii*) defines the active power generated by the PV source connected at node R-16 by reading stored data from experimental measurements, and (*iv*) defines the on/off status of the load breaker at node R-18.
- InverterA_controller and InverterB_controller contain the control loops of the EPCs connected at node R16 and R10, respectively. The control implements the structure shown in Figure 5 and imposes the active and reactive power injection issued by the central microgrid controller and received and dispatched in the real-time simulation by means of the block Central_ctr_interaction.
- PQ-calculator computes the per-phase active and reactive power measured at each node of the network.
- RMS-calculator computes the per-phase rms voltage measured at each node of the network.
- PV_R16 generates the current reference for the PV source on the basis of the power profile generation recorded from a real plant and accessed inside the block Central_ctr_interaction.
- Analog out routes the real-time simulation signals of the model to the analog output ports of the digital simulation platform, whose outputs are then measured by the DL850EV.

The devised partition allows to perform simulation with small discretization steps on FPGA for those models presenting fast dynamics and allow the execution of even complex control algorithm on CPU. The accuracy of the developed real-time simulation setup has been verified by comparison with equivalent desktop computer simulation models executed with variable time step simulations.

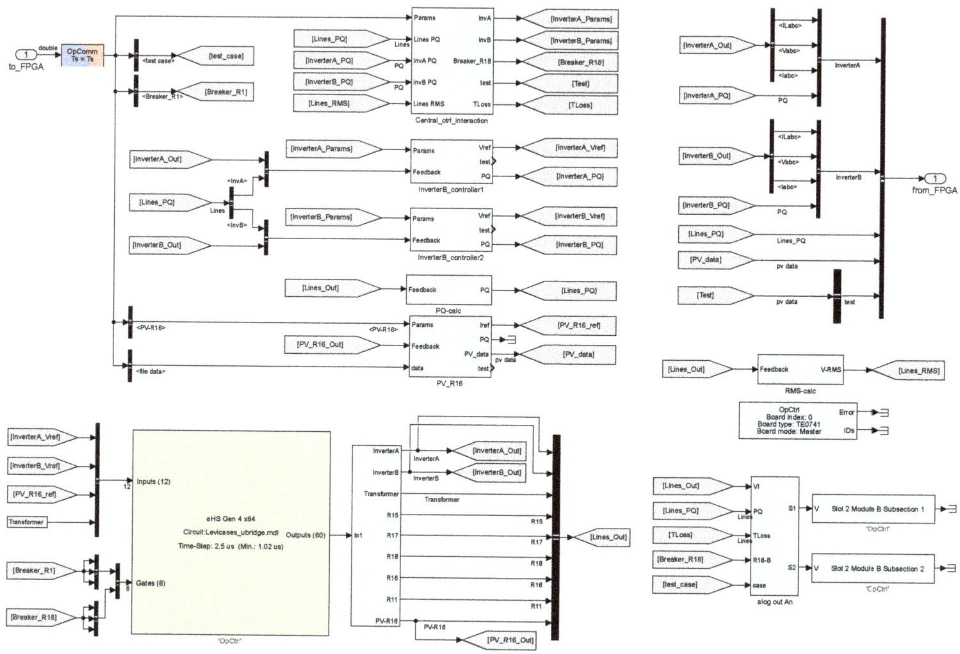

Figure 10. System model in the real-time simulator. The green bottom-left block, executed on FPGA, models the distribution network and the EPCs hardware, the other blocks, executed on CPU, model the EPC controls shown in Figures 4 and 5.

Table 1. Electronic power processor (EPC) parameters.

Parameter	Symbol	Value
Output filter inductor	L_o	1 mH
Inductor ESR	R_L	15 mΩ
Output filter capacitance	C_o	220 µF
Capacitor ESR	R_C	10 mΩ
Voltage regulator proportional gain	k_{pv}	$3.3 \times 10^{-2}\,\mathrm{V}^{-1}$
Voltage regulator integral gain	k_{iv}	$508\,\mathrm{s}^{-1}\mathrm{V}^{-1}$
Current regulator proportional gain	k_{pi}	$4.5\,\mathrm{A}^{-1}$
Current regulator integral gain	k_{ii}	$2715\,\mathrm{s}^{-1}\mathrm{A}^{-1}$
Droop $P-f$ coefficient	k_p	$3.037 \times 10^{-3}\,\mathrm{Hz/W}$
Droop $P-f$ integral gain	h_p	1
Droop $Q-V$ coefficient	k_q	$8.478 \times 10^{-5}\,\mathrm{V/VAr}$
Droop $Q-V$ integral gain	h_q	300

4.2. Results

The control system has been tested in different operating conditions. Six scenarios of operation with different kinds of constraints while minimizing the cost function (11) are described next.

- Case 0, no control: the situation in which distributed EPCs are switched off while loads and sources exchange nominal active and reactive powers as defined in Figure 7.
- Case 1, reactive power control: distributed EPCs are controlled by the power flow controller to generate only reactive power in order to minimize the cost (11).
- Case 2, active and reactive power control: as in the previous case, but with active power control too.
- Case 3, power balance at PCC (point of common coupling): distributed EPCs are controlled by the power flow controller to generate the active and reactive power needed to balance among the phases the power absorption at the PCC while keeping the value of the total power exchange at the PCC as in Case 0.
- Case 4, autonomous operation: distributed EPCs are controlled to generate the active and reactive power that is needed to achieve per-phase zero power flow at the PCC, satisfying, in this way, the whole power needs of the microgrid.
- Case 5, demand–response at PCC: distributed EPCs are controlled to generate the active and reactive power to achieve balanced and purely active power flow at the PCC. The active power reference is set to 10 kW.

The obtained results in steady-state conditions are reported in Table 2, while transient behaviors in relevant conditions are displayed in Figure 11. As a general remark, it is possible to note that voltage deviations and distribution loss significantly reduce in all the considered test cases when distributed EPCs are active and controlled by the microgrid controller. Each operating condition is considered in more details in the following.

Table 2. Simulation results.

Test Case	Maximum Voltage Deviation (V)	Total P Fed by Sources (kW)	Total Q Fed by Sources (kVAr)	Total P Absorbed by Loads (kW)	Total Q Absorbed by Loads (kVAr)	P Fed at PCC (kW)	Q Fed at PCC (kVAr)	Distrib. Losses (W)
Case 0	4.6	46.25	28.68	45.60	28.19	46.246	28.68	647
Case 1	2.3	46.77	28.95	46.24	28.55	46.464	4.088	530
Case 2	0.9	47.06	29.07	46.85	28.93	7.028	3.694	210
Case 3	3.0	46.35	28.73	45.79	28.27	46.683	28.975	560
Case 4	1.2	47.29	29.20	47.05	29.05	0.045	−0.04	237
Case 5	1.3	47.13	29.11	46.91	28.96	10.09	0.07	220

Figure 11a refers to the transition from Case 0 to Case 1 and then to Case 2. It displays, from top to bottom, the total active and reactive power exchanged at the PCC and the measured total distribution losses. Case 1 allows a significant improvement in terms of power quality, indeed the maximum voltage deviation of the network nodes halves and the power factor at the PCC increases from 0.850 to 0.996. In addition, distribution loss is mitigated by 18%. Significant improvements are obtained by activating active power control too: in this case distribution loss decreases by an additional 60% in the last time interval of Figure 11a.

Figure 11b refers to Case 3, reporting the currents through the three phases at the PCC and the correspondingly measured active and reactive powers. Noticeably, power balance is achieved accurately, being the active and reactive powers among the phases at the PCC are of equal amount.

The dynamics related to power flow control are important when considering demand–response at the point of connection with the main grid, which is demonstrated in Figure 11c,d. The former relates to a constraint of zero active and reactive power exchange, the latter to a constraint of pure, balanced active power exchanged with the main grid. Additionally in this case, the optimal coordination of distributed EPCs allows to accurately track the given references.

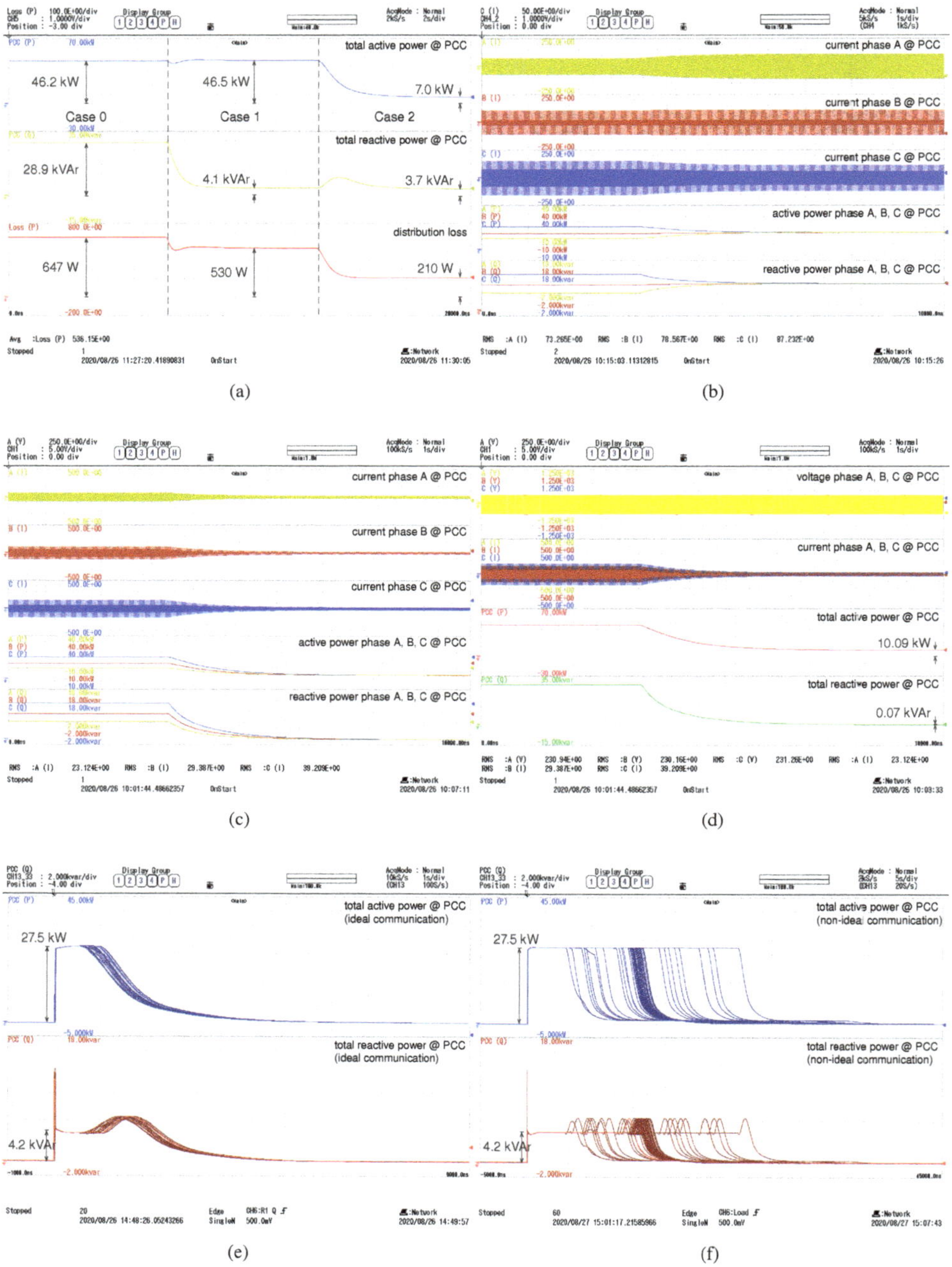

Figure 11. Results in the considered operating conditions. (**a**) Distribution losses in Case 0, Case 1, and Case 2; (**b**) active and reactive power balancing at the PCC (point of common coupling) in Case 3; (**c**) zero active and reactive power reference at the PCC in Case 4; (**d**) demand–response at the PCC with requested active power equal to 10 kW and zero reactive power in Case 5; (**e**) disturbance rejection at the PCC after a load change within the grid with ideal communication; (**f**) as in the previous case but with non-ideal communication.

The impact on the dynamic performances of including communication impairments in the system is considered too in the validation. Figure 11e–f show the effect seen at the PCC after a sudden increase of power absorption by 27.5 kW due to the connection of the load at node R18. During the condition in which the control system is set to impose zero power flow at PCC, Figure 11e refers to the case of ideal communication whilst Figure 11f refers to non-ideal and impaired communication with 20% packet loss and random latency in the interval [300 ms, 1 s]. Considering the random nature of the considered aspect, a batch of sixty consecutive acquisitions are simultaneously reported in the figures. Figure 11e shows that the control brings back to zero the controlled quantity in a time compatible with the chosen control frequency of 2 Hz if the communication is ideal. Instead, in the case of Figure 11e, dynamics are significantly delayed even though steady-state performance is preserved. Such kind of considerations are important in the design of master–slave microgrid architectures (see, e.g., [31,32]) where a single EPC is expected to buffer possible energy unbalances. In such a case, Figure 11e indicates that a master EPC should be able to buffer about 82.5 kJ, while in case of communication fault as in Figure 11e the buffered energy increases to 137.5 kJ, which corresponds to 40% and 68%, respectively, of the capacity of a super-capacitor energy storage as in [33]. In addition, it is possible to note an oscillation due to the active and reactive coupling of the droop control loop.

Figure 12 shows a long-term simulation over ten hours, in which the distribution loss in case of communication impairments is specifically considered. The simulation comprises variability in load power absorption at node R18, which is periodically switched on and off, and in PV power generation, sampled with a time step of one second, which considers a measured profile during a cloudy day. The simulation is run with ideal as well as non-ideal communication, showing negligible impact of communication impairments on power flow optimization: distribution loss increases from an average of 349 W to an average of 354 W, which confirms the effectiveness of the approach.

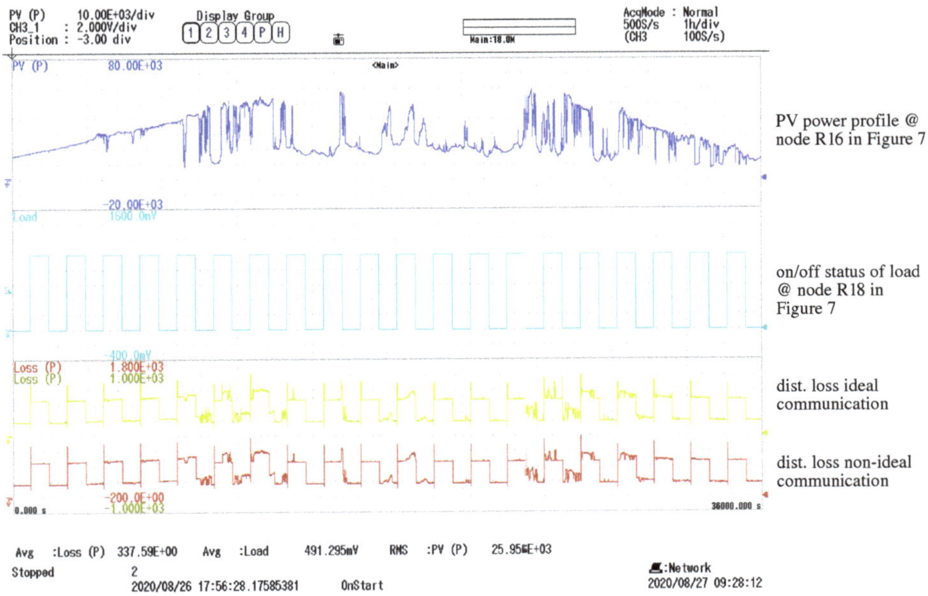

Figure 12. Long-term simulation (i.e., 10 h) without/with communication impairments.

Finally, Figure 13 shows a long-term simulation over twenty hours in which the load at node R15 is switched on/off randomly, the load at node R18 absorbs the actual power profile measured in a subsection of a university campus, and the source at node R16 generates the actual power profile measured at a PV installation. The figure shows, specifically, the instantaneous distribution loss when the distributed converters are disabled (i.e., case without optimal control) and the case in which the converters are controlled according to the presented optimal power flow (i.e., case with optimal control). Notably, the average distribution loss amounts to 1.122 kW in the first case, while, enabling distributed converters to respond to the optimal control signals, the distribution loss decreases to 392 W, which corresponds to a distribution loss reduction of 65%.

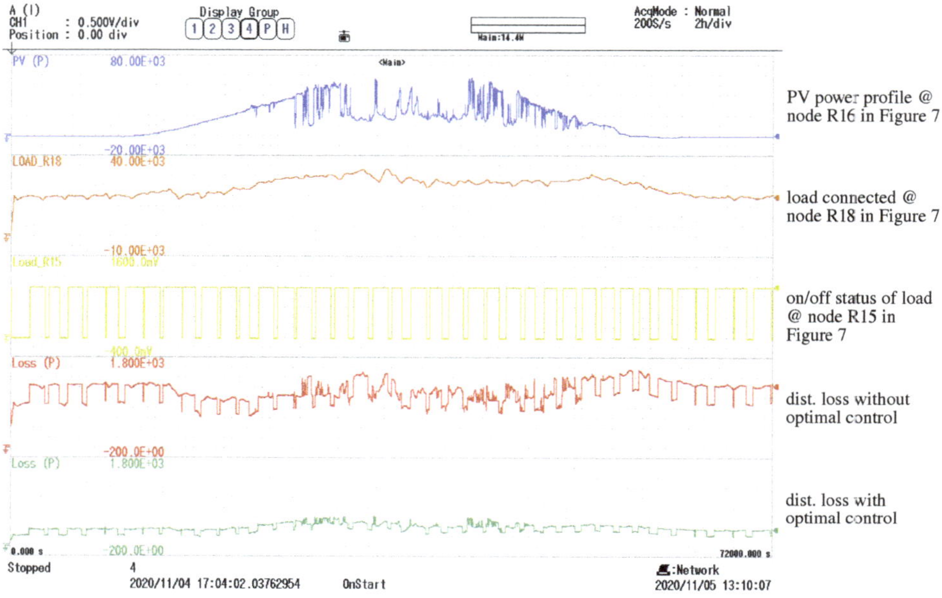

Figure 13. Long-term simulation (i.e., 20 h) without/with optimal control.

In summary, the described control approach is based on a general algorithm with an explicit solution of the control problem. In this way, it is not affected by convergence issues even in case of communication failures and it is adaptable to generic networks and operating conditions. The approach is validated by means of a real-time simulation setup that is described in detail herein. This setup can be considered for the validation of generic systems involving fast electronic power converters, control algorithms for management of distributed resources, and a communication infrastructure allowing data exchange for distributed resources coordination. The considered scenario and validation testbed are useful in the forthcoming power-electronics-dominated grids, where control and communication play an important and substantial role [34,35].

5. Conclusions

In this paper, an optimal power flow control method for microgrids and its real-time validation considering a benchmark low-voltage distribution network with distributed energy resources has been presented. Distributed resources are considered interfaced to the network by means of electronic power converters implementing a specific power-based droop controller. Such a controller allows power-flow control when operating connected to the main grid, while preserving the capability of operating islanded in case of accidental disconnection. The power-flow control method is implemented centrally at microgrid level and set to dispatch power references to the distributed electronic converters. The reported results show that the local control of distributed converters driven by the control signals computed by the described power flow control method achieves minimum distribution losses, improved power quality indices, and fulfillment of constraints at the point of connection with the main grid. The real-time experimental setup allowed to investigate steady-state operation as well as short-term and long-term dynamics in realistic generation and communication network conditions.

The reported results show the effectiveness of the described power flow control in several conditions of practical interest. Notably, the approach is suitable to provide optimal coordination of distributed energy resources to respond, for example, to demand–response requests issued by entities at higher layers in the power-system control hierarchy. The described real-time simulation testbench may be taken as reference for other studies concerning power electronics nominated grids exploiting communication for distributed resources coordination. Future studies may regard the overall operation of a grid subsection that integrate the shown dynamic power optimization and long-term energy optimizations. Actually, the control approach was devised to provide the flexibility needed to comply, in the future, with the European vision where clusters of prosumers aggregated in microgrids will actively participate to the electrical market by trading their energy resources.

Author Contributions: Conceptualization, P.M. and T.C.; methodology, P.T.; software, P.T., H.A.; validation, H.A., P.T.; formal analysis, P.T.; data curation H.A.; investigation, H.A., T.C.; writing—original draft preparation, T.C.; writing—review and editing, P.T.; visualization, H.A., T.C.; supervision, P.M., P.T., T.C.; resources T.C.; project administration, T.C.; funding acquisition, P.T., P.M., T.C. All authors have read and agreed to the published version of the manuscript.

Funding: This research was mainly funded by Interdepartmental Centre Giorgio Levi Cases, project NEBULE. Part of the funding is also coming form the PRIN project HEROGRIDS.

Conflicts of Interest: The authors declare no conflict of interest. The funders had no role in the design of the study; in the collection, analyses, or interpretation of data; in the writing of the manuscript, or in the decision to publish the results.

References

1. Simmini, F.; Agostini, M.; Coppo, M.; Caldognetto, T.; Cervi, A.; Lain, F.; Carli, R.; Turri, R.; Tenti, P. Leveraging Demand Flexibility by Exploiting Prosumer Response to Price Signals in Microgrids. *Energies* **2020**, *13*, 3078. [CrossRef]
2. Gough, M.; Santos, S.F.; Javadi, M.; Castro, R.; Catalão, J.P.S. Prosumer Flexibility: A Comprehensive State-of-the-Art Review and Scientometric Analysis. *Energies* **2020**, *13*, 2710. [CrossRef]
3. Publications Office of the EU. *Clean Energy for All Europeans*; European Union: Luxembourg, 2019; p. 24.
4. Tenti, P.; Caldognetto, T. On Microgrid Evolution to Local Area Energy Network (E-LAN). *IEEE Trans. Smart Grid* **2019**, *10*, 1567–1576. [CrossRef]
5. *IEEE Standard for Interconnection and Interoperability of Distributed Energy Resources with Associated Electric Power Systems Interfaces*; IEEE Std 1547-2018 (Revision of IEEE Std 1547-2003); IEEE: Piscataway, NJ, USA, 2018; pp. 1–138.
6. Dambrauskas, P.; Syed, M.; Blair, S.; Irvine, J.; Abdulhadi, I.; Burt, G.; Bondy, D.E.M. Impact of realistic communications for fast-acting demand side management. *CIRED Open Access Proc. J.* **2017**, *2017*, 1813–1817. [CrossRef]
7. Ly, A.; Bashash, S. Fast Transactive Control for Frequency Regulation in Smart Grids with Demand Response and Energy Storage. *Energies* **2020**, *13*, 4771. [CrossRef]
8. Cha, H.J.; Won, D.J.; Kim, S.H.; Chung, I.Y.; Han, B.M. Multi-Agent System-Based Microgrid Operation Strategy for Demand Response. *Energies* **2015**, *8*, 14272–14286. [CrossRef]
9. Silva, V.A.; Aoki, A.R.; Lambert-Torres, G. Optimal Day-Ahead Scheduling of Microgrids with Battery Energy Storage System. *Energies* **2020**, *13*, 5188. [CrossRef]
10. Elkazaz, M.; Sumner, M.; Pholboon, S.; Davies, R.; Thomas, D. Performance Assessment of an Energy Management System for a Home Microgrid with PV Generation. *Energies* **2020**, *13*, 3436. [CrossRef]
11. Ochoa, L.F.; Padilha-Feltrin, A.; Harrison, G.P. Evaluating distributed generation impacts with a multiobjective index. *IEEE Trans. Power Deliv.* **2006**, *21*, 1452–1458. [CrossRef]
12. Sao, C.K.; Lehn, P.W. Control and Power Management of Converter Fed Microgrids. *IEEE Trans. Power Syst.* **2008**, *23*, 1088–1098. [CrossRef]
13. Panjaitan, S.D.; Kurnianto, R.; Sanjaya, B.W. Flexible Power-Sharing Control for Inverters-Based Microgrid Systems. *IEEE Access* **2020**, *8*, 177984–177994. [CrossRef]

14. Tayab, U.B.; Roslan, M.A.B.; Hwai, L.J.; Kashif, M. A review of droop control techniques for microgrid. *Renew. Sustain. Energy Rev.* **2017**, *76*, 717–727. [CrossRef]
15. Tian, W.; Zhang, Y.; Fu, R.; Zhao, Y.; Wang, G.; Winter, R. Modeling and control architecture of source and load management in islanded power systems. In Proceedings of the 2015 IEEE Energy Conversion Congress and Exposition (ECCE), Montreal, QC, Canada, 20–24 September 2015; pp. 3407–3413. [CrossRef]
16. Summers, A.; Johnson, J.; Darbali-Zamora, R.; Hansen, C.; Anandan, J.; Showalter, C. A Comparison of DER Voltage Regulation Technologies Using Real-Time Simulations. *Energies* **2020**, *13*. [CrossRef]
17. Barragán-Villarejo, M.; García-López, F.D.P.; Marano-Marcolini, A.; Maza-Ortega, J.M. Power System Hardware in the Loop (PSHIL): A Holistic Testing Approach for Smart Grid Technologies. *Energies* **2020**, *13*, 3858. [CrossRef]
18. Estrada, L.; Vázquez, N.; Vaquero, J.; de Castro, Á.; Arau, J. Real-Time Hardware in the Loop Simulation Methodology for Power Converters Using LabVIEW FPGA. *Energies* **2020**, *13*, 373. [CrossRef]
19. Caldognetto, T.; Buso, S.; Mattavelli, P. Digital Controller Development Methodology Based on Real-Time Simulations with LabVIEW FPGAc Hardware-Software Toolset. *Electronics* **2013**, *17*. [CrossRef]
20. Buso, S.; Caldognetto, T. Rapid Prototyping of Digital Controllers for Microgrid Inverters. *IEEE J. Emerg. Sel. Top. Power Electron.* **2015**, *3*, 440–450. [CrossRef]
21. D'Agostino, F.; Kaza, D.; Martelli, M.; Schiapparelli, G.P.; Silvestro, F.; Soldano, C. Development of a Multiphysics Real-Time Simulator for Model-Based Design of a DC Shipboard Microgrid. *Energies* **2020**, *13*, 3580. [CrossRef]
22. Montoya, J.; Brandl, R.; Vishwanath, K.; Johnson, J.; Darbali-Zamora, R.; Summers, A.; Hashimoto, J.; Kikusato, H.; Ustun, T.S.; Ninad, N.; et al. Advanced Laboratory Testing Methods Using Real-Time Simulation and Hardware-in-the-Loop Techniques: A Survey of Smart Grid International Research Facility Network Activities. *Energies* **2020**, *13*, 3267. [CrossRef]
23. Oh, S.J.; Yoo, C.H.; Chung, I.Y.; Won, D.J. Hardware-in-the-Loop Simulation of Distributed Intelligent Energy Management System for Microgrids. *Energies* **2013**, *6*, 3263–3283. [CrossRef]
24. Erseghe, T.; Lorenzon, F.; Tomasin, S.; Costabeber, A.; Tenti, P. Distance measurement over PLC for dynamic grid mapping of smart micro grids. In Proceedings of the 2011 IEEE International Conference on Smart Grid Communications (SmartGridComm), Brussels, Belgium, 17–20 October 2011; pp. 487–492. [CrossRef]
25. Bolognani, S. Chapter 13—Grid Topology Identification via Distributed Statistical Hypothesis Testing. In *Big Data Application in Power Systems*; Arghandeh, R., Zhou, Y., Eds.; Elsevier: Amsterdam, The Netherlands, 2018; pp. 281–301. [CrossRef]
26. Hoffmann, N.; Fuchs, F.W. Minimal Invasive Equivalent Grid Impedance Estimation in Inductive–Resistive Power Networks Using Extended Kalman Filter. *IEEE Trans. Power Electron.* **2014**, *29*, 631–641. [CrossRef]
27. Rocabert, J.; Luna, A.; Blaabjerg, F.; Rodríguez, P. Control of Power Converters in AC Microgrids. *IEEE Trans. Power Electron.* **2012**, *27*, 4734–4749. [CrossRef]
28. Bidram, A.; Davoudi, A. Hierarchical Structure of Microgrids Control System. *IEEE Trans. Smart Grid* **2012**, *3*, 1963–1976. [CrossRef]
29. Lissandron, S.; Mattavelli, P. A controller for the smooth transition from grid-connected to autonomous operation mode. In Proceedings of the 2014 IEEE Energy Conversion Congress and Exposition (ECCE), Pittsburgh, PA, USA, 14–18 September 2014; pp. 4298–4305.
30. CIGRE Task Force C6.04.02. *Benchmark Systems for Network Integration of Renewable and Distributed Energy Resources*; Technical Report; International Council on Large Electric Systems: Paris, France, 2014.
31. Alfergani, A.; Khalil, A. Modeling and control of master-slave microgrid with communication delay. In Proceedings of the 2017 8th International Renewable Energy Congress (IREC), Amman, Jordan, 21–23 March 2017; pp. 1–6.
32. Caldognetto, T.; Tenti, P. Microgrids Operation Based on Master–Slave Cooperative Control. *IEEE J. Emerg. Sel. Top. Power Electron.* **2014**, *2*, 1081–1088. [CrossRef]
33. Maxwell Technologies. *BMOD0130 P056 B03—56-V Module Supercapacitor Datasheet*; Maxwell Technologies: San Diego, CA, USA, 2020.

34. Burgos, R.; Sun, J. The Future of Control and Communication: Power Electronics-Enabled Power Grids. *IEEE Power Electron. Mag.* **2020**, *7*, 34–36. [CrossRef]
35. Tavassoli, B.; Fereidunian, A.; Mehdi, S. Communication system effects on the secondary control performance in microgrids. *IET Renew. Power Gener.* **2020**, *14*, 2047–2057. [CrossRef]

Publisher's Note: MDPI stays neutral with regard to jurisdictional claims in published maps and institutional affiliations.

energies

MDPI

Article

Effect of Daily Forecasting Frequency on Rolling-Horizon-Based EMS Reducing Electrical Demand Uncertainty in Microgrids

Giuseppe La Tona, Maria Carmela Di Piazza * and Massimiliano Luna

Consiglio Nazionale delle Ricerche (CNR), Istituto di Ingegneria del Mare (INM), via Ugo La Malfa 153, 90146 Palermo, Italy; giuseppe.latona@cnr.it (G.L.T.); massimiliano.luna@cnr.it (M.L.)
* Correspondence: mariacarmela.dipiazza@cnr.it

Abstract: Accurate forecasting is a crucial task for energy management systems (EMSs) used in microgrids. Despite forecasting models destined to EMSs having been largely investigated, the analysis of criteria for the practical execution of this task, in the framework of an energy management algorithm, has not been properly investigated yet. On such a basis, this paper aims at exploring the effect of daily forecasting frequency on the performance of rolling-horizon EMSs devised to reduce demand uncertainty in microgrids by adhering to a reference planned profile. Specifically, the performance of a sample EMS, where the forecasting task is committed to a nonlinear autoregressive network with exogenous inputs (NARX) artificial neural network (ANN), has been studied under different daily forecasting frequencies, revealing a representative trend relating the forecasting execution frequency in the EMS and the reduction of uncertainty in the electrical demand. On the basis of such a trend, it is possible to establish how often is convenient to repeat the forecasting task for obtaining increasing performance of the EMS. The obtained results have been generalized by extending the analysis to different test scenarios, whose results have been found coherent with the identified trend.

Keywords: energy management system; forecasting error; rolling horizon; demand uncertainty; microgrids

Citation: La Tona, G.; Di Piazza, M.C.; Luna, M. Effect of Daily Forecasting Frequency on Rolling-Horizon-Based EMS Reducing Electrical Demand Uncertainty in Microgrids. *Energies* **2021**, *14*, 1598. https://doi.org/10.3390/en14061598

Academic Editor: Gianfranco Chicco

Received: 18 February 2021
Accepted: 9 March 2021
Published: 13 March 2021

1. Introduction

The microgrid paradigm has gained interest in the last decade as a promising solution for a progressive decarbonization of the energy mix and a more efficient, flexible, and economic operation of electrical power systems [1]. Similarly, the concept of a building-integrated microgrid has emerged due to a twofold reason: firstly, for environmental issues, since the electrical power consumption of buildings is about 32% of the total production worldwide with a contribution to the global greenhouse gas (GHG) emissions approximately equal to 30%; secondly, for the growing influence of building-integrated grid-connected renewables on power quality and stability of electrical distribution grids [2,3].

Moreover, the recent provisions introduced for the grid-connected renewable generators encourage the consumers to switch from the role of passive energy users to the role of active energy producers; thus, the users contribute with energy supply and ancillary power quality services to the main power grid, according to the concept of the prosumer microgrid, where a prosumer is a user who can both produce and consume the energy [4,5].

On such a basis, numerous technical contributions have been proposed in the literature on energy management systems (EMSs) for residential/commercial microgrids encompassing renewable generators and battery storage systems (BSS), with the aim of improving energy efficiency and reducing the energy bill by means of demand response (DR) or alternative optimization-based strategies [6–8]. Such EMSs, if properly coordinated with the power grid upstream, can also produce benefits to the main grid manager; for example, by limiting the peak of energy demand and the power loss on electrical feeders and by preventing adverse events, such as poor renewable-based power production, energy price fluctuations, voltage limits breaches, and so on [4,6,9–11]. Therefore, EMSs are considered nowadays a relevant technical solution for the enhancement of the efficiency, reliability, and economy of smart microgrids [12–14].

EMSs are increasingly gaining interest also in the field of vehicular electrical systems. Recently, the use machine learning techniques for energy management with a special focus on thermal and battery degradation issues has been proposed. In [15,16] reinforcement learning strategies (a soft actor-critic deep reinforcement learning strategy and the deep deterministic policy gradient algorithm combined with an expert-assistance system, respectively) are used to get optimal allocation of power in hybrid electric buses; furthermore, thermal and battery degradation issues are considered in the energy management algorithm formulation, obtaining improved performance in terms of training efforts, optimization, and overheat protection with respect to existing strategies.

With regard to microgrids, forecasting of generation and load demand is one of the most important tools for an EMS—it is usually employed to plan and schedule optimal power flows in a microgrid and has a significant impact on the effectiveness and performance of the EMS [17]. Several energy-related data forecasting models have been proposed in technical literature, ranging from simple persistence methods, to physical models, to time series linear models, up to artificial intelligence (AI)-based models, including machine learning-based approaches such as the most recent solutions based on deep learning [18,19]. Among them, the most robust and reliable methods, according to the literature, are those based on AI, e.g., artificial neural networks (ANNs), where a suitable trade-off between simplicity and accuracy is also taken into account as an enabling factor for a straightforward real-world implementation of forecasting functionalities within EMSs [20].

As far as the use of forecasting techniques within EMSs is concerned, some authors have focused their interest on establishing the most performing/convenient algorithms to forecast the energy-related variables involved in grid energy management. For example, in [5], a new hybrid machine learning-based method is developed to precisely and simultaneously forecast the microgrid's variables with the aim of improving a DR-based operation; this method encompasses an adaptive neuro-fuzzy inference system, a multilayer perceptron ANN, and a radial basis function ANN.

In [21], some possible candidate forecasting methods have been investigated with regard to their convenience and cost efficiency, rather than to their accuracy, identifying simple forecasting models, based on regression analyses using linear, seasonal linear, and quadratic formulations.

Integration and coordination issues between forecasters and optimizers in EMSs have been explored as well. In [22], for instance, a fully automated control in a home energy management system (HEMS) including all parts of the smart microgrid architecture (non-invasive load identification, forecasting, optimization, renewable energy sources, and storage elements) is proposed; the work is particularly focused on the coordinated use of a forecasting model based on the long short-term memory (LSTM) and an optimization strategy based on a genetic algorithm (GA), working respectively on the prediction and optimal scheduling of load demand.

Other recent contributions in technical literature have been focused on the evaluation of forecasting errors on the EMS performance. A HEMS minimizing cost and energy loss, and improving self-consumption is presented in [23]; here, the accuracy of several forecasting methods is evaluated, demonstrating that forecasting errors in both load demand and renewable generation produce adverse effects on the HEMS performance and, thus, on the household energy cost and lost energy.

Still in the context of the evaluation of the effects of forecasting errors, ref. [24] analyzes the sensitivity of a residential microgrid's battery control to load demand and generation forecasting errors when the battery is involved in arbitrage for residential consumers under the Time of Use pricing scheme; it is demonstrated that low forecasting accuracy has an impact in terms of energy losses.

Although the investigation of the most performing or convenient forecasting techniques for EMSs has received a considerable deal of attention in recent studies, the selection criterion of the daily forecasting frequency, in the framework of an energy management algorithm, has not been properly investigated yet. Specifically, defining how many times the generation/load forecasting should be updated in the typical 24 h horizon of an EMS to achieve enhanced performance is still an open question.

To start filling this gap, in this paper, the effect of forecasting task execution frequency has been evaluated considering the rolling horizon EMS presented in [11]. Such an EMS relies on a two-stage algorithm designed to minimize the electrical demand uncertainty in a grid-connected building microgrid by trying to adhere to a reference planned profile of the power exchanged with the upstream grid. The EMS under consideration was tested considering as a case study a microgrid comprising photovoltaic (PV) generation, battery storage, and electrical loads. The forecasting of both PV-generated power and load demand within the considered EMS is executed once a day according to the multi-step-ahead approach, and it is based on the nonlinear autoregressive network with exogenous inputs (NARX) ANN, which is a structure realizing a good balance between performance and simplicity.

In this paper, different update rates of the forecasting task performed using the NARX ANN have been introduced in the EMS execution, and their effect on the demand uncertainty obtained as a result of the EMS operation has been quantitatively evaluated. The analysis has revealed a clear trend relating the forecasting execution frequency in the EMS and the reduction of uncertainty in the electrical demand. Specifically, a reduction of the demand uncertainty is observed for increased forecasting execution frequency up to 48 times a day. Such a trend also allows for the establishment of to what an extent is convenient to repeat the forecasting task in the considered case study for obtaining still significant improvements of the EMS performance. To generalize the observed behavior, the same experiments have been carried out considering also synthetic forecasted profiles and artificial reduction of forecasting errors during the day according to a model. These additional results have confirmed the identified trend. It is worth observing that the presented analysis results could be useful to designers and practitioners involved in the design of EMSs for microgrids.

The paper is organized as follows. In Section 2, the main features and the operating mode of the two-stage rolling horizon EMS developed in [11] are recalled. Section 3 presents the case study, i.e., the characteristics of the building's microgrid and the used datasets of PV power generation and load demand. Section 4 deals with the daily forecasting frequency analysis, including a comparison between the results obtained by the EMS in the real case and in the generalized cases. Finally, a discussion of the results is presented, and conclusions are drawn.

2. Fundamentals of the Considered EMS

As stated in Section 1, the paper aims at evaluating the effect of daily forecasting frequency on the performance of EMSs that aim at reducing demand uncertainty in a grid-connected building microgrid trying to adhere to a reference planned profile of the power exchanged with the main power grid. Such a reference profile can be obtained in different ways depending on the goal to be pursued. For example, it can be computed aiming to achieve the minimum operating cost of a microgrid [11,25], possibly considering hourly variable prices or other DR management techniques enforced by the distribution system operator (DSO) [26,27]. In other cases, the reference profile is obtained, reshaping the expected profile of the consumed or injected power at the point of common coupling; for example, reducing its peak-to-mean ratio to avoid incurring in curtailment measures applied by the DSO to preserve grid stability [28,29]. As a third example, it is possible to follow an approach aimed at planning the usage of predictable loads and managing unpredictable loads in real time thanks to a PV-battery backup system to ensure a reliable and efficient power supply in countries where regular and frequent blackouts occur in the main grid [30].

The proposed investigation will be performed referring to the EMS presented in [11], whose specific main features are described hereinafter; however, the obtained results are representative of the behavior of many EMSs that aim at reducing demand uncertainty trying to adhere to a reference power profile.

The considered EMS pursues two goals at the same time thanks to a two-stage operation aiming at providing advantages both to the end user and to the DSO by computing optimal power references for the controllers of all the electrical devices of the microgrid (i.e., local renewable or non-renewable generators, storage systems, electrical loads as well as the connection to the main grid).

In the first stage, i.e., Planning, it forecasts the 24 h ahead profiles of load demand and environmental variables tied to renewable generation based on past data using an ANN. To achieve good results, the forecasting is performed using the NARX ANN, which has been successfully used in time-series modeling thanks to its simple implementation and its adaptive learning process, even with small-scale data [31]. In the considered case, the exogenous input is the environmental temperature for both PV generation and load demand forecasting. In particular, as far as the dependence of electrical power demand on environmental temperature is concerned, it should be observed that a power consumption scenario framed in a low carbon vision is considered by means of massive use of electrification, e.g., heat pumps for heat, ventilation, and air conditioning (HVAC); on such a basis, a strong dependence of load demand on temperature is envisaged. Based on the forecast data, the EMS solves an optimization problem to find the optimal 24 h profiles of the power flows for all the electrical devices of the microgrid that allow the minimization of the user's cash flow in the whole next day. The point-by-point sum of such optimal profiles gives the planned profile of the power exchanged with the grid in the next 24 h or, for short, the planned GEPP (grid-exchanged power profile), which is a vector of power values across the time steps of a day (positive when the main grid supplies the microgrid, negative when an excess of power from renewable generators is injected into the main grid). The planned GEPP is transmitted to the DSO over a secure Internet connection as an obligation to which the user commits themself. Being based on forecast data, the planned GEPP could be affected by forecasting errors.

In the second stage of the EMS, an Online Replanning task is repeated at each time step of the day using measured data and aiming at minimizing the deviation between the actual GEPP and the planned one that was transmitted to the DSO the day before. In this way, besides minimizing the user's cash flow, the considered EMS also reduces the uncertainty on the GEPP against forecasting errors on the load demand and renewable generation profiles. The DSO can take advantage of this last feature and, by exploiting also the natural statistical compensation of users' deviations, can rely on a quite predictable aggregate demand profile to optimize power dispatching and improve its economic planning policy [32]. It should be observed that the goal of minimizing the cash flow is in any case pursued in compliance with the enforced physical, technical, and contractual constraints. Therefore, considering a more extended electrical system governed by the EMS does not imply criticalities for the DSO.

The Online Replanning task is based on solving another optimization problem, and it is not greedy because it does not aim at minimizing only the deviation at the current time step. Instead, it performs an optimization also considering the future time steps of the current day in a rolling-horizon fashion. Therefore, at each execution of the Online Replanning task, the first set of power references for the microgrid devices, i.e., the set computed for the current time step, is actually sent to the device controllers via an internal communication network. Instead, the other sets of power references, i.e., those computed for the future time steps, are discarded. After the execution of the Online Replanning task in the last time step of the day, a new cycle restarts from the Planning stage.

As explained, in the original formulation of [11], the forecasting task is performed only once in a day (e.g., at midnight); in particular, the whole 24 h ahead forecasted profiles are used in the Planning stage, whereas at each execution of the Online Replanning stage a smaller and smaller residual portion of the same profiles is used to consider future time steps in a rolling-horizon fashion. In Section 4, instead, the new approach will be presented. Using such an approach, the forecasting task is performed several times in a day so that, at each execution of the Online Replanning stage, the residual portions of the most recent forecasted profiles are considered, always in a rolling-horizon fashion. The Planning stage, instead, is kept unchanged and processes the first forecasted profile.

Optimization Problems Solved in the Considered EMS and Forecasting Method

The variables considered in the optimization problems solved in the Planning and Online Replanning tasks are the power flows between couples of microgrid devices and the battery state of charge (SOC).

The objective function of the optimization problem of the Planning task is the cash flow (CF) of the end user, which can be expressed as:

$$f_{obj}(x) = \sum_{t=1}^{N} CF_t \tag{1}$$

$$CF_t = \begin{cases} GEPP_t \Delta_t C_{buy}, & GEPP_t \geq 0 \\ GEPP_t \, \Delta_t C_{sell}, & otherwise \end{cases} \tag{2}$$

In the previous equation, C_{sell} and C_{buy} represent the cost of sold/purchased energy, respectively; Δ_t is the time step duration; N is the total number of time steps in a day; and $t \in [1, N]$ represents the time step index. The EMS tries to minimize the objective function (1) while satisfying the following set of physical and design constraints at each time step:

1. Each variable should have a non-negative value.
2. Power balance must be respected at each node.
3. The maximum contractual limit for the power exchanged with the main grid must be enforced.
4. The maximum limits for battery charging/discharging power must be enforced.
5. The battery SOC must not exceed the maximum and minimum limits recommended by the manufacturer.
6. The SOC difference between successive time steps is computed according to battery power flow, battery efficiency, and time step duration.
7. The starting SOC of a day is the final SOC of the previous day.
8. The SOC profile should be cyclic between consecutive days.

All the above constraints are easily translated into equality or inequality relationships among the variables of the system, i.e., the power flows between couples of microgrid devices and the battery SOC. Their mathematical formulation is reported in [11].

The solution of the optimization problem expressed by (1), (2), and the eight constraints is a vector of N sets of power references for the microgrid devices (i.e., one set for each time step of the next day) that imply the minimum cash flow for the end user in the whole next day. As stated before, computing the algebraic sum of the power references at each time step gives the planned GEPP, which is transmitted to the DSO. However, since forecasting is never perfect, the actual profiles of load demand and renewable generation of the next day will be different from those considered in the Planning stage. Therefore, the actual GEPP will be different to a greater or lesser extent with respect to the planned GEPP due to forecasting errors.

Aiming to reduce the deviation of the actual GEPP against the self-committed one, at each time step of the day the Online Replanning task recomputes the set of power references for the microgrid devices based on the measured instantaneous values of load demand, renewable generation, and battery SOC. It is worth observing that, in our study, battery SOC values are assumed to be measurable by dedicated subsystems. Actually, technical literature proposes numerous model-based SOC estimation techniques, such as electrochemical (ECM)-based observers; particularly, extensive discussions on the recursive Bayesian filters, including the Kalman filter, the extended Kalman filter, the unscented Kalman filter, particle filter, and other adaptive filters, such as the proportional-integral observer, the sliding mode observer, and the nonlinear predictive filter, have been proposed for the online estimation of battery SOC [33].

It is worth noting that a greedy minimization of the instantaneous differences between the actual GEPP against the self-committed one could lead to having a completely full or empty battery unexpectedly during the day. In such a circumstance, the buffering behavior of the battery would become unidirectional (i.e., only for charging or discharging); thus, the degree of freedom that is requested to null or minimize the GEPP deviation could potentially be lost resulting in very high and prolonged errors against the self-committed GEPP.

Therefore, the algorithm of the Online Replanning task proceeds considering the remaining duration of the day in a rolling horizon fashion and computes at each execution the set of power references not only for the current time step $t = i$ (using measured values) but also for the following steps $t \in [i+1, N]$ (using forecast values). However, only the reference power values for the current time step ($t = i$) are sent to the controllers of the microgrid devices, whereas the other values (for $t > i$) are discarded. Hence, the objective function that is minimized in the Online Replanning task is the maximum absolute difference between the self-committed GEPP at time step t (i.e., $\overline{GEPP_t}$) and the actual GEPP (i.e., $GEPP_t$):

$$f_{obj}(x) = \max_{i \le t \le N} \left| \overline{GEPP_t} - GEPP_t \right|. \tag{3}$$

The constraints for (3) are easily obtained by modifying those of the Planning stage in the following way:

- The time intervals on which the constraints are defined are progressively changed at each execution in a rolling-horizon fashion.
- Power balance constraints are split to account for measured values (at $t = i$) and for forecast values (at $t > i$).

The defined optimization problems are non-linear since they involve piecewise-linear functions. They have been solved using either mixed-integer linear programming [11] (desktop PC implementation) or dynamic programming [32] (embedded implementation); this last method does not imply dependencies from third-party solvers; therefore, it is most suited to be implemented in embedded processing platforms.

As explained before, the NARX ANN is used to perform the forecasting of PV generation and load demand. It predicts the next values of a time series (e.g., load demand and solar irradiance) as a function of the observed past values of the time series and of the observed values of another variable (i.e., the exogenous input). The analytical formulation of such a function relies on the following equation:

$$y_{t+i} = f(x_{t+o}, \ldots, x_t, \ldots, x_{t-d}, y_t, y_{t-1} \ldots, y_{t-d}), \; i \in [1, o], \tag{4}$$

where y and x are, respectively, the variable to be forecast and the exogenous input of the ANN considered at a discrete time step, d is the delay, and o is the number of output units. The (vectorial) function f is approximated by a multilayer perceptron and, to forecast future values, the output is fed back to the network as input. More specifically, data windows from the forecasted series and from the exogenous series are fed, through a delay operator, to the input units of the ANN, which outputs a prediction for the next o time steps. These, in turn, are fed back to the ANN as input recursively to forecast for the desired number of future time steps. Figure 1 depicts the structure of the NARX ANN.

Further details on the structure definition and training process related to the NARX ANN are given in [32]. The hyperparameters have been reported in Section 3.

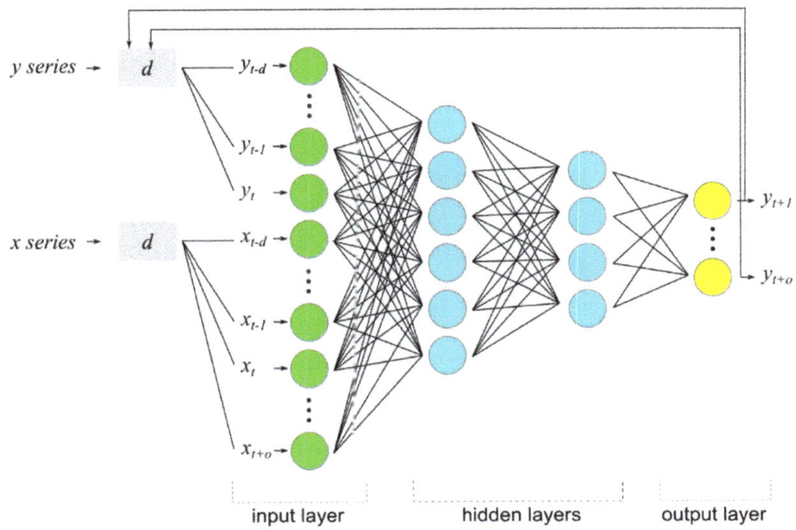

Figure 1. Scheme of the used configuration of the nonlinear autoregressive network with exogenous inputs artificial neural network (NARX ANN).

A previously developed forecasting module [20], based on the NARX ANN and using the TensorFlow library [34], has been used to provide updated forecasted profiles at the user's request. It is worth underlining that the forecasting module only manages the forecasting phase, whereas the NARX ANN's training is executed offline. Thanks to this feature, the forecasting task can be embedded in online calculations. Furthermore, the forecasting module has been extended to also provide profiles generated according to user-specified error formulations to allow for investigation of the different scenarios described in Section 4.

3. Case Study

With no loss of generality, the analysis of the daily forecasting frequency was performed on the EMS managing a single household encompassing a direct current (DC) microgrid consisting of a PV generator, a set of electrical loads considered as a single aggregate load, a BSS, and a connection to the utility grid at the point of common coupling (PCC). As shown in Figure 2, each of these elements is electrically connected to the DC bus through a dedicated power electronic converter. Furthermore, each of these converters is coupled with a smart sensor, which is capable of sending detected data and receiving actuation commands [35]. The EMS and the forecasting module communicate with the smart sensors to manage the household grid and to store historical data. Finally, the EMS communicates with the Utility/DSO to send the self-committed GEPP.

The system considered for the study is further defined by the parameters of the EMS and of the considered energy market, reported in Table 1, and by the parameters of the electrical system reported in Table 2.

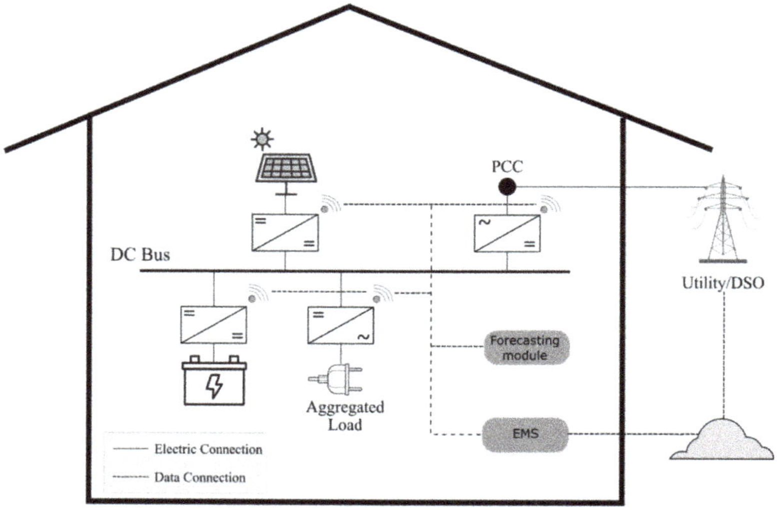

Figure 2. Representation of the elements of the household considered in the study.

Table 1. Parameters of the energy management system (EMS) and of the energy market.

Parameter	Value	Description
Δ_t	30 min	Duration of EMS time step
N	48	Number of time steps in planning horizon
SOC_{tol}	15%	Tolerance on end of day SOC constraint
SOC_{end_min}	85%	Minimum end of day SOC
SOC_{end_max}	45%	Maximum end of day SOC
C_{buy}	0.1024 (EUR/kWh)	Price of purchased energy
C_{sell}	0.0850 (EUR/kWh)	Price of sold energy

Table 2. Electrical system parameters.

Parameter	Value	Description
C_b	6 kWh	BSS capacity
η	0.95	BSS efficiency
SOC_{min}	20%	Lower bound for BSS SOC
SOC_{max}	100%	Upper bound for BSS SOC
P_{xb}	6 kW	Maximum charge/discharge power for BSS
P_{xg}	6 kW	Maximum grid contractual power
P_{xp}	6 kW	Maximum power of the PV generator

Simulations for the different considered scenarios were performed considering three publicly available datasets as measured data for the EMS, each providing a different measurement: (1) electric power consumption, (2) solar irradiance, and (3) temperature. The first dataset provides measurements of electric power consumption in one household located in Sceaux, France, with a one-minute sampling rate over a period of almost 4 years [36,37]. The second dataset contains solar irradiance measurements for the same location with a time step of one minute obtained from the Copernicus Atmosphere Monitoring Service (CAMS) radiation service [38,39]. Finally, the third dataset holds temperature measurements for the same location at one-hour intervals obtained from the Photovoltaic Geographical Information System (PVGIS) service [40]. All the datasets were resampled at a 30 min rate.

The NARX ANN used in this study for solar irradiance forecasting has an input delay equal to 7, 2 hidden layers with 20 and 10 neurons respectively and an output consisting of 1 neuron (it predicts one future time step at a time and needs 48 feedback iterations to forecast one day). As for the load demand forecasting, the used NARX ANN has an input delay equal to 7, 2 hidden layers with 30 and 20 neurons respectively and an output consisting of 12 neurons (it predicts 12 future time steps at a time, and it needs 4 feedback iterations to forecast one day).

4. Analysis and Results

In this section, the analysis of the effect of daily forecasting frequency on EMS performance is presented and discussed. Starting from preliminary considerations about the dependence of forecasting performance on execution starting time during the day, some test scenarios have been defined. In particular, three alternative formulations of forecasting error trend have been proposed, against which the proposed method was compared and validated. The considered error metrics on the GEPP are the normalized maximum error (NMAX) and the normalized root mean square error (NRMSE). Both errors are normalized with respect to the maximum grid contractual power P_{xg}.

Intuitively, forecasting performance improves as new information on the considered variable is available during a day. For example, in the case of solar irradiance, as the hours of the day progress, it is clearer whether it is a cloudy or a sunny day. This intuition has been verified by forecasting the solar irradiance using the NARX ANN starting at different hours of the day and terminating at midnight (i.e., with a reducing forecasting window). As an example, Figure 3 shows such forecasted profiles for a sample day in the case of solar irradiance, and it is possible to notice that these profiles are less dispersed and closer to the measured profile when forecasting starts in the later hours. In quantitative terms, the results for solar irradiance, particularly in terms of NMAX, are shown by the metrics reported in Table 3.

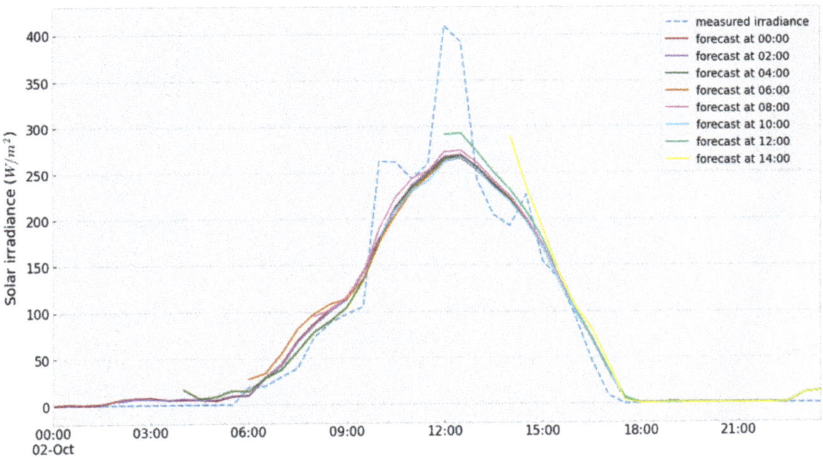

Figure 3. Solar irradiance forecasted profiles during a sample day, starting at successive two-hour intervals.

Table 3. Error metrics for solar irradiance forecasting starting at successive two-hour intervals for a sample day.

Forecasting at Hour	00:00	02:00	04:00	06:00	08:00	10:00	12:00	14:00
NMAX	25.06%	25.10%	25.37%	25.84%	24.09%	25.99%	20.81%	16.86%

On the basis of such considerations, EMS simulations have been executed for a 50-day period with different daily forecasting frequency using the same datasets and starting with an initial SOC of 80%. Specifically, the considered daily forecasting frequencies are {1, 2, 4, 8, 12, 24, 48} times a day, and four test scenarios have been considered:

- Scenario 1: real forecasting;
- Scenario 2: modified real forecasting;
- Scenario 3: high-performance generated forecasting;
- Scenario 4: average-performance generated forecasting.

The first scenario aims at evaluating the effect of different daily forecasting frequencies on EMS performance in the real application, i.e., when the EMS uses the NARX ANN-based forecasting at each invocation. The other scenarios have been devised aiming at generalizing the results obtained in Scenario 1 considering different degrees of abstraction in the formulation of forecasting errors. In particular, two aspects have been considered; on the one hand, Scenario 2 still uses the NARX ANN-based forecasting, whereas in Scenarios 3 and 4, synthetic forecasted profiles with given error metrics have been generated as described in the following section to unlink the test results from the specific forecasting method; furthermore, Scenarios 2, 3, and 4 use forecasting errors that are artificially reduced at successive invocations according to the previously verified intuition (see Figure 3 and Table 3) as described in the following section.

4.1. Generation of Synthetic Forecasted Profiles and Artificial Reduction of Error

The synthetic forecasted profiles considered in Scenario 3 (high-performance generated forecasting) and in Scenario 4 (average-performance generated forecasting) are obtained, adding to the measured profiles m an error modeled as the sum of a Gaussian term and an impulsive term in order to impose predefined NRMSE (10% in Scenario 3 and 15% in Scenario 4) and NMAX (50% in Scenario 3 and 100% in Scenario 4) metrics. More specifically, the Gaussian term has zero mean and a standard deviation calculated as

$$\sigma = NRMSE[\max(m(t_i)) - \min(m(t_i))], \tag{5}$$

where $NRMSE$ is the desired value of this metric. Instead, the impulsive term is calculated by sampling from a uniform distribution over the time steps of the day and adding at such steps the value of the desired maximum absolute error:

$$e = NMAX[\max(m(t_i)) - \min(m(t_i))], \tag{6}$$

where $NMAX$ is the desired value of this metric.

The resulting profile is clipped to zero for negative values to maintain the physical meaning of the profile. For this reason and since the pseudo random samplings are executed on a relatively small number of samples (i.e., the time steps of a day), the resulting values of the metrics slightly deviate from the desired values but nonetheless they are still a good approximation.

In Scenarios 2 to 4, the forecasting error is reduced at successive invocations within a day by progressively reducing a parameter of the forecasting error model proposed in [41]. According to such an approach, the forecasting module produces (for both solar irradiance and load demand) a forecasting once a day and artificially revises it during the day as a weighted sum of the measured profile and the initial forecasting, according to

$$F^{rev}(t_i) = m(t_i) + \lambda(F(t_i) - m(t_i)),\ \lambda \in [0,1],\ i \in \{0,1,2,\ldots,N\}, \tag{7}$$

where F is the initial forecasting, m is the measured profile, F^{rev} is the revised forecasting, and λ is a weight coefficient (for $\lambda = 0$ the revised forecasting equals the measured profile, for $\lambda = 1$ the revised forecasting equals the initial forecasting). In our approach, the weight λ is linearly reduced during the day as

$$\lambda = \frac{1}{(t_0 - t_n)}(t_i - t_0) + 1. \tag{8}$$

In this way, the forecasting accuracy is progressively increased.

4.2. Simulation Results and Discussion

Simulation results for Scenario 1 (real forecasting) are shown in Figure 4. It is clear that the EMS performance in terms of GEPP errors improves with the increase of the daily forecasting frequency. In particular, a reduction of 5.3% on NMAX is observed and a slight reduction of about 0.2% is noted on NRMSE.

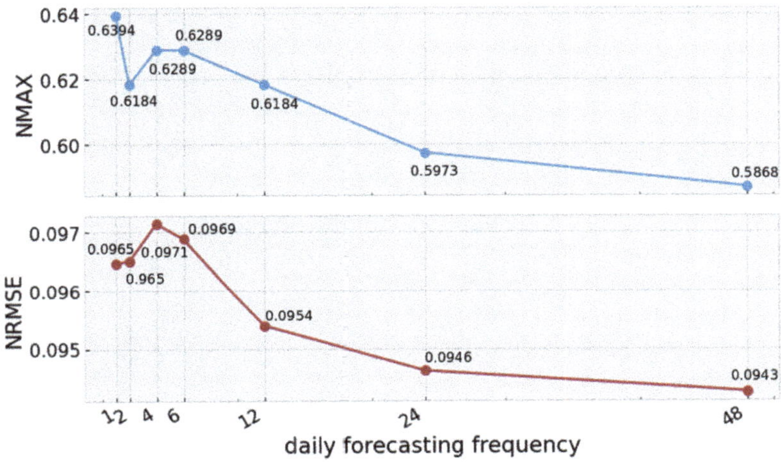

Figure 4. Error metrics on the grid-exchanged power profile (GEPP) for different values of daily forecasting frequency in Scenario 1.

It is worth noting that the EMS shows a good capability in reducing the errors on the GEPP (EMS output). Additionally, for the sake of completeness, the error metrics of the forecasted load and generation profiles (EMS inputs) are given in Table 4.

Table 4. Total errors of load and generation forecasted profiles as seen by the planning stage of the EMS in Scenarios 1 and 2.

	Load	Generation
NRMSE	13.33%	9.40%
NMAX	65.52%	45.7%

The GEPP metrics reported in Figure 4 are calculated over the entire 50-day simulation period; however, it is also useful to observe how the daily error evolves. As an example, Figure 5 shows the daily GEPP NMAX values for the different daily forecasting frequencies; it is possible to note that in some days the error metric reduction with increased forecasting frequency is more significant than in others.

A similar trend, i.e., error reduction for increasing daily forecasting frequency, has been obtained performing simulations in Scenario 2. In fact, the decreasing trend is clearly shown in Figure 6 and GEPP NMAX is reduced up to 16.9%, whereas GEPP NRMSE is reduced up to 1.7%.

Simulation results in Scenario 3 with an imposed forecasting NMAX of 50% and an imposed forecasting NRMSE of 10%, show a decrease of GEPP NMAX up to 13.8% and of GEPP NRMSE up to 1.2%, as depicted in Figure 7.

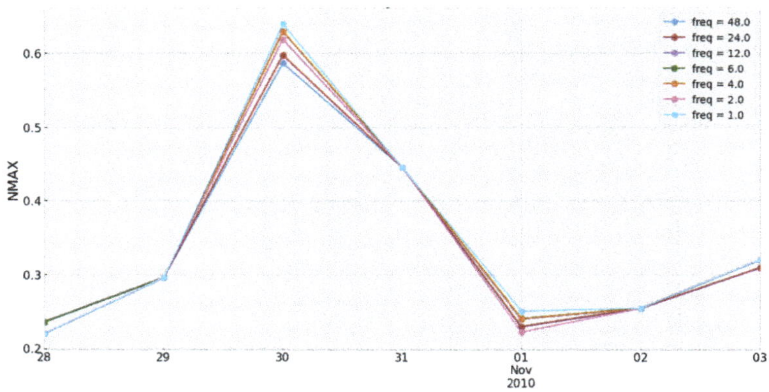

Figure 5. Daily normalized maximum error (NMAX) of GEPP on a sample week for different daily forecasting frequencies in Scenario 1.

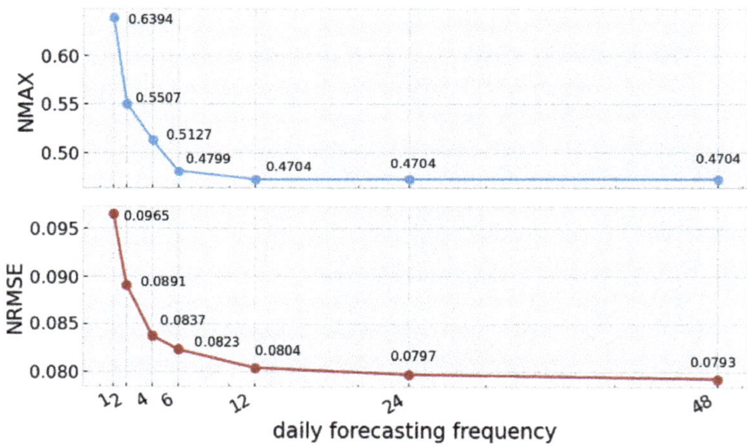

Figure 6. Error metrics on GEPP for different values of daily forecasting frequency in Scenario 2.

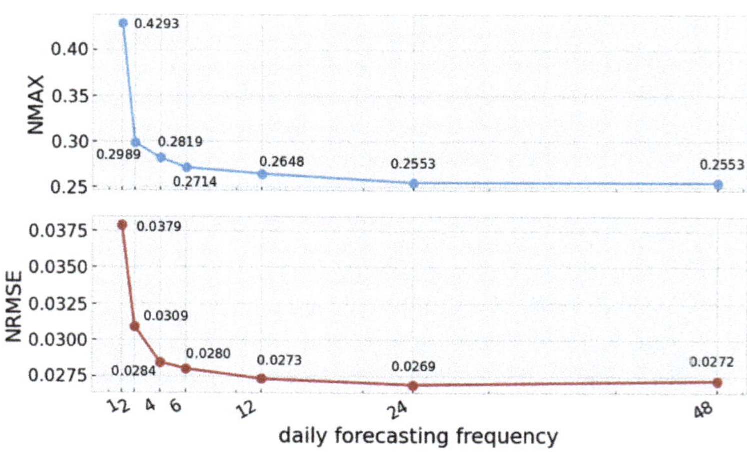

Figure 7. Error metrics on GEPP for different values of daily forecasting frequency in Scenario 3.

Finally, simulation results in Scenario 4, with an imposed forecasting NMAX of 100% and an imposed forecasting NRMSE of 15%, confirm the trend of performance increase with the increase of the daily forecasting frequency as Figure 8 shows. In particular, NMAX decreases up to 24.6% and NRMSE decreases up to 1.9%.

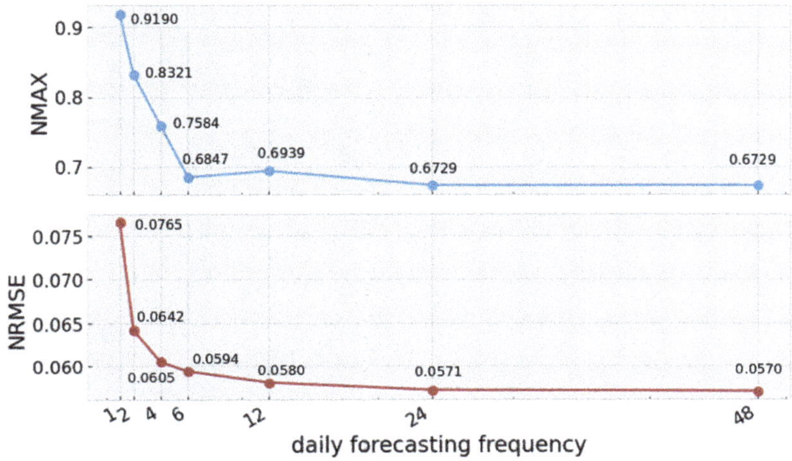

Figure 8. Error metrics on GEPP for different values of daily forecasting frequency in Scenario 4.

The simulation results in all the considered scenarios show the same trend with an improvement of EMS performance with the increase of daily forecasting frequency up to 48 times a day. The observed decreasing trends could be expected based on the intuition verified with the help of Figure 3 and Table 3. However, thanks to the performed investigation, it has been found that the trends present a saturation for higher frequency values. In particular, it can be observed that in Scenarios 2 to 4 for frequencies higher than or equal to 12 times a day, the metrics almost flatten. On the other hand, in Scenario 1, the decrease of the metrics is more gradual and between the three higher frequencies considered there is an NMAX difference of 3.2%. This can be explained considering that this scenario uses real forecasting whose errors only qualitatively follow the ideal model of forecasting error decrease during the day.

Given the presence of a saturation in the error metric trends, an attempt was made to describe them by usual saturated mathematical laws, such as exponential or hyperbolic functions. However, such functions are not suitable to fit the observed trends. Furthermore, the error metric trends exhibit some outliers due to one or both the following reasons—the presence of stochastic variables (forecasting errors) and the use of nonlinear optimization algorithms. These outliers (observed in Scenario 1 and Scenario 4) do not hinder the validity of the general findings.

It is worth noting that the most significant observed result is the reduction of the NMAX metric. This implies that a more frequent forecasting execution improves the EMS capability to handle generation/demand peak errors. On the other hand, a lower but still appreciable reduction of NRMSE is also exhibited in all the scenarios.

Comparing the results obtained in the different scenarios, it is possible to observe that an increase of the daily forecasting frequency leads to a more significant reduction of the error metrics when the less performing forecasting is involved, e.g., in Scenario 4.

The consistency of the obtained results in all the considered scenarios is a key achievement of the proposed analysis since it proves the general validity of such results, under the assumption that forecasting of considered variables improves as the time of the day progresses.

Furthermore, the repetition of the forecasting task usually implies a negligible computational cost for the execution of EMS steps. As a matter of fact, in the considered case study, the forecasting task takes less than a second to produce a result, whereas the duration of the EMS time step is much higher, i.e., 30 min. Therefore, it can be executed even at each time step without impairing the EMS operation. However, in the worst case, i.e., for a very large electrical network and a very complex forecasting algorithm, the total EMS computation time could increase beyond the EMS time step; in such a case, it would not be feasible to repeat the forecasting at each time step and a tradeoff between lower forecasting frequency and higher EMS performance would be needed. Nonetheless, according to the observed trends, a modest decrease of the daily forecasting frequency with respect to once per time step would still provide a significant reduction of the error metrics.

5. Conclusions

Accurate forecasting of load demand and renewable generation is a pivotal element of optimization-based EMSs in microgrids. For such a reason, beside the refinement of forecasting models, investigation on how and how often to run forecasting task within an EMS deserves attention.

This paper aims at analyzing the effect of daily forecasting frequency on the performance of EMSs used to reduce demand uncertainty in microgrids by adhering to a reference planned profile. Specifically, the paper considers a sample EMS of this kind, shows the results obtained using different daily forecasting frequencies, and generalizes such results, considering also synthetic forecasted profiles and artificial reduction of forecasting errors during the day according to a model.

In summary, the analysis showed that: (1) a more frequent forecasting execution improves the EMS capability to handle generation/demand peak errors, thus reducing NMAX error of the grid-exchanged power profile (theoretically up to 24.6%); (2) the reduction of the NRMSE is lower but still appreciable (theoretically up to about 2%); (3) the accuracy of the forecasting method affects the entity of the error metric reduction obtained with a more frequent forecasting execution; (4) the higher the daily forecasting frequency, the smaller the entity of the progressive error reduction (saturation effect).

The above considerations highlight the importance of using an accurate forecasting method and executing the forecasting task several times in a day in order to achieve the highest performance of the EMS. Finally, the repetition of such a task usually implies a negligible increase of the computational cost of the EMS algorithm while improving the EMS performance; otherwise, for very large electrical networks and very complex forecasting algorithms, a suitable compromise could be chosen by the EMS designer.

Author Contributions: Conceptualization, G.L.T., M.C.D.P., and M.L.; methodology, G.L.T., M.C.D.P., and M.L.; software, G.L.T.; validation, G.L.T., M.C.D.P., and M.L.; writing—original draft preparation, G.L.T., M.C.D.P., and M.L.; writing—review and editing, G.L.T., M.C.D.P., and M.L.; supervision, M.L.; funding acquisition, G.L.T. All authors have read and agreed to the published version of the manuscript.

Funding: This research was funded by Italian Ministry of University and Research (MUR) grant number 2017WA5ZT3_004, program PRIN2017, and project HEROGRIDS.

Institutional Review Board Statement: Not applicable.

Informed Consent Statement: Not applicable.

Data Availability Statement: Not applicable.

Conflicts of Interest: The authors declare no conflict of interest.

References

1. Sabzehgar, R. A Review of AC/DC Microgrid—Developments, Technologies, and Challenges. In Proceedings of the 2015 IEEE Green Energy and Systems Conference (IGESC), Long Beach, CA, USA, 9 November 2015; pp. 11–17.
2. Liu, N.; Chen, Q.; Liu, J.; Lu, X.; Li, P.; Lei, J.; Zhang, J. A Heuristic Operation Strategy for Commercial Building Microgrids Containing EVs and PV System. *IEEE Trans. Ind. Electron.* **2015**, *62*, 2560–2570. [CrossRef]
3. Lee, D.; Cheng, C.C. Energy Savings by Energy Management Systems: A Review. *Renew. Sustain. Energy Rev.* **2016**, *56*, 760–777. [CrossRef]
4. Jafari, M.; Malekjamshidi, Z.; Zhu, J.; Khooban, M.H. A Novel Predictive Fuzzy Logic-Based Energy Management System for Grid-Connected and Off-Grid Operation of Residential Smart Microgrids. *IEEE J. Emerg. Sel. Top. Power Electron.* **2020**, *8*, 1391–1404. [CrossRef]
5. Faraji, J.; Ketabi, A.; Hashemi-Dezaki, H.; Shafie-Khah, M.; Catalao, J.P.S. Optimal Day-Ahead Self-Scheduling and Operation of Prosumer Microgrids Using Hybrid Machine Learning-Based Weather and Load Forecasting. *IEEE Access* **2020**, *8*, 157284–157305. [CrossRef]
6. Pinzon, J.A.; Vergara, P.P.; Da Silva, L.C.P.; Rider, M.J. Optimal Management of Energy Consumption and Comfort for Smart Buildings Operating in a Microgrid. *IEEE Trans. Smart Grid* **2019**, *10*, 3236–3247. [CrossRef]
7. Hao, H.; Corbin, C.D.; Kalsi, K.; Pratt, R.G. Transactive Control of Commercial Buildings for Demand Response. *IEEE Trans. Power Syst.* **2017**, *32*, 774–783. [CrossRef]
8. Deng, S.; Wang, R.Z.; Dai, Y.J. How to Evaluate Performance of Net Zero Energy Building—A Literature Research. *Energy* **2014**, *71*, 1–16. [CrossRef]
9. Rahimi, F.; Ipakchi, A. Using a Transactive Energy Framework. *IEEE Electrif. Mag.* **2016**, *4*, 23–29. [CrossRef]
10. Samad, T.; Koch, E.; Stluka, P. Automated Demand Response for Smart Buildings and Microgrids: The State of the Practice and Research Challenges. *Proc. IEEE* **2016**, *104*, 726–744. [CrossRef]
11. Di Piazza, M.C.; La Tona, G.; Luna, M.; Di Piazza, A. A Two-Stage Energy Management System for Smart Buildings Reducing the Impact of Demand Uncertainty. *Energy Build.* **2017**, *139*, 1–9. [CrossRef]
12. Di Piazza, A.; Di Piazza, M.C.; La Tona, G.; Luna, M. An Artificial Neural Network-Based Forecasting Model of Energy-Related Time Series for Electrical Grid Management. *Math. Comput. Simul.* **2021**, *184*, 294–305. [CrossRef]
13. Diagne, M.; David, M.; Lauret, P.; Boland, J.; Schmutz, N. Review of Solar Irradiance Forecasting Methods and a Proposition for Small-Scale Insular Grids. *Renew. Sustain. Energy Rev.* **2013**, *27*, 65–76. [CrossRef]
14. Di Piazza, A.; Di Piazza, M.C.; Ragusa, A.; Vitale, G. Environmental Data Processing by Clustering Methods for Energy Forecast and Planning. *Renew. Energy* **2011**, *36*, 1063–1074. [CrossRef]
15. Wu, J.; Wei, Z.; Li, W.; Wang, Y.; Li, Y.; Sauer, D. Battery Thermal- and Health-Constrained Energy Management for Hybrid Electric Bus Based on Soft Actor-Critic DRL Algorithm. *IEEE Trans. Ind. Inform.* **2020**, 3203. [CrossRef]
16. Wu, J.; Wei, Z.; Liu, K.; Quan, Z.; Li, Y. Battery-Involved Energy Management for Hybrid Electric Bus Based on Expert-Assistance Deep Deterministic Policy Gradient Algorithm. *IEEE Trans. Veh. Technol.* **2020**, *69*, 12786–12796. [CrossRef]
17. Panapongpakorn, T.; Banjerdpongchai, D. Short-Term Load Forecast for Energy Management Systems Using Time Series Analysis and Neural Network Method with Average True Range. In Proceedings of the 2019 1st International Symposium on Instrumentation, Control, Artificial Intelligence, and Robotics, ICA-SYMP, Bangkok, Thailand, 16–18 January 2019; pp. 86–89. [CrossRef]
18. Yildiz, B.; Bilbao, J.I.; Sproul, A.B. A Review and Analysis of Regression and Machine Learning Models on Commercial Building Electricity Load Forecasting. *Renew. Sustain. Energy Rev.* **2017**, *73*, 1104–1122. [CrossRef]
19. Le, T.; Vo, M.T.; Vo, B.; Hwang, E.; Rho, S.; Baik, S.W. Improving Electric Energy Consumption Prediction Using CNN and Bi-LSTM. *Appl. Sci.* **2019**, *9*, 4237. [CrossRef]
20. La Tona, G.; Luna, M.; Di Piazza, A.; Di Piazza, M.C. Development of a Forecasting Module Based on Tensorflow for Use in Energy Management Systems. In Proceedings of the IECON 2019 45th Annual Conference of the IEEE Industrial Electronics Society, Lisbon, Portugal, 14–19 September 2019; pp. 3063–3068. [CrossRef]
21. Lee, D. Low-Cost and Simple Short-Term Load Forecasting for Energy Management Systems in Small and Middle-Sized Office Buildings. *Energy Explor. Exploit.* **2020**. [CrossRef]
22. Shakir, M.; Biletskiy, Y. Forecasting and Optimisation for Microgrid in Home Energy Management Systems. *IET Gener. Transm. Distrib.* **2020**, *14*, 3458–3468. [CrossRef]
23. Elkazaz, M.; Sumner, M.; Pholboon, S.; Davies, R.; Thomas, D. Performance Assessment of an Energy Management System for a Home Microgrid with PV Generation. *Energies* **2020**, *130*, 3436. [CrossRef]
24. Kiedanski, D.; Hashmi, M.U.; Busic, A.; Kofman, D. Sensitivity to Forecast Errors in Energy Storage Arbitrage for Residential Consumers. In Proceedings of the 2019 IEEE International Conference on Communications, Control, and Computing Technologies for Smart Grids, SmartGridComm 2019, Beijing, China, 21–23 October 2019. [CrossRef]
25. Yaghmaee Moghaddam, M.H.; Leon-Garcia, A. A Fog-Based Internet of Energy Architecture for Transactive Energy Management Systems. *IEEE Internet Things J.* **2018**, *5*, 1055–1069. [CrossRef]
26. Li, W.; Yuen, C.; Hassan, N.U.; Tushar, W.; Wen, C.; Wood, K.L.; Hu, K.; Liu, X. Demand Response Management for Residential Smart Grid: From Theory to Practice. *IEEE Access* **2015**, *3*, 2431–2440. [CrossRef]

27. Paterakis, N.G.; Erdinç, O.; Bakirtzis, A.G.; Catalão, J.P.S. Optimal Household Appliances Scheduling under Day-Ahead Pricing and Load-Shaping Demand Response Strategies. *IEEE Trans. Ind. Inform.* **2015**, *11*, 1509–1519. [CrossRef]
28. Di Piazza, M.C.; Luna, M.; La Tona, G.; Di Piazza, A. Improving Grid Integration of Hybrid PV-Storage Systems Through a Suitable Energy Management Strategy. *IEEE Trans. Ind. Appl.* **2019**, *55*, 60–68. [CrossRef]
29. Zhao, Z.; Lee, W.C.; Shin, Y.; Song, K.-B. An Optimal Power Scheduling Method for Demand Response in Home Energy Management System. *IEEE Trans. Smart Grid* **2013**, *4*, 1391–1400. [CrossRef]
30. Khoury, J.; Mbayed, R.; Salloum, G.; Monmasson, E. Design and Implementation of a Real Time Demand Side Management under Intermittent Primary Energy Source Conditions with a PV-Battery Backup System. *Energy Build.* **2016**, *133*, 122–130. [CrossRef]
31. Siegelmann, H.T.; Horne, B.G.; Giles, C.L. Computational Capabilities of Recurrent NARX Neural Networks. *IEEE Trans. Syst. Man Cybern. Part B* **1997**, *27*, 208–215. [CrossRef] [PubMed]
32. La Tona, G.; Luna, M.; Di Piazza, A.; Di Piazza, M.C. Towards the Real-World Deployment of a Smart Home EMS: A DP Implementation on the Raspberry Pi. *Appl. Sci.* **2019**, *9*, 2120. [CrossRef]
33. Wei, Z.; Zhao, J.; He, H.; Ding, G.; Cui, H.; Liu, L. Future Smart Battery and Management: Advanced Sensing from External to Embedded Multi-Dimensional Measurement. *J. Power Source* **2021**, *489*, 229462. [CrossRef]
34. Abadi, M.; Barham, P.; Chen, J.; Chen, Z.; Davis, A.; Dean, J.; Devin, M.; Ghemawat, S.; Irving, G.; Isard, M.; et al. TensorFlow: Large-Scale Machine Learning on Heterogeneous Systems. *arXiv* **2015**, arXiv:1603.04467.
35. Luna, M.; La Tona, G.; Di Piazza, M.C.; Pucci, M.; Accetta, A.; Taibi, D.; Vetro, C.; La Grassa, R. A Prototype of Wireless Sensor for Data Acquisition in Energy Management Systems. In Proceedings of the 2018 IEEE International Conference on Environment and Electrical Engineering and 2018 IEEE Industrial and Commercial Power Systems Europe (EEEIC/I&CPS Europe), Palermo, Italy, 12–15 July 2018. [CrossRef]
36. Hebrail, G.; Berards, A. Individual Household Electric Power Consumption Data Set. UCI Machine Learning Repository. Available online: https://archive.ics.uci.edu/ml/datasets/individual+household+electric+power+consumption (accessed on 11 February 2021).
37. Dua, D.; Graff, C. UCI Machine Learning Repository. 2017. Available online: http://archive.ics.uci.edu/ml (accessed on 31 January 2021).
38. Lefèvre, M.; Oumbe, A.; Blanc, P.; Espinar, B.; Gschwind, B.; Qu, Z.; Wald, L.; Schroedter-Homscheidt, M.; Hoyer-Klick, C.; Arola, A.; et al. McClear: A New Model Estimating Downwelling Solar Radiation at Ground Level in Clear-Sky Conditions. *Atmos. Meas. Tech.* **2013**, *6*, 2403–2418. [CrossRef]
39. Gschwind, B.; Wald, L.; Blanc, P.; Lefèvre, M.; Schroedter-Homscheidt, M.; Arola, A. Improving the McClear Model Estimating the Downwelling Solar Radiation at Ground Level in Cloud-Free Conditions—McClear-v3. *Meteorol. Z.* **2019**, *28*, 147–163. [CrossRef]
40. Huld, T.; Müller, R.; Gambardella, A. A New Solar Radiation Database for Estimating PV Performance in Europe and Africa. *Sol. Energy* **2012**, *86*, 1803–1815. [CrossRef]
41. Hobbs, B.F.; Jitprapaikulsarn, S.; Konda, S.; Chankong, V.; Loparo, K.A.; Maratukulam, D.J. Analysis of the Value for Unit Commitment of Improved Load Forecasts. *IEEE Trans. Power Syst.* **1999**, *14*, 1342–1348. [CrossRef]

energies

Article

A Comparison of Time-Domain Implementation Methods for Fractional-Order Battery Impedance Models

Brian Ospina Agudelo [1,2,*], Walter Zamboni [1,*] and Eric Monmasson [2]

[1] Dipartimento di Ingegneria dell'Informazione ed Elettrica e Matematica Applicata DIEM, Università degli Studi di Salerno, 84084 Fisciano, Italy
[2] Laboratoire SATIE, CY Cergy Paris Université, 95000 Neuville-sur-Oise, France; eric.monmasson@u-cergy.fr
* Correspondence: bospinaagudelo@unisa.it (B.O.A.); wzamboni@unisa.it (W.Z.)

Abstract: This paper is a comparative study of the multiple RC, Oustaloup and Grünwald–Letnikov approaches for time domain implementations of fractional-order battery models. The comparisons are made in terms of accuracy, computational burden and suitability for the identification of impedance parameters from time-domain measurements. The study was performed in a simulation framework and focused on a set of ZARC elements, representing the middle frequency range of Li-ion batteries' impedance. It was found that the multiple RC approach offers the best accuracy–complexity compromise, making it the most interesting approach for real-time battery simulation applications. As for applications requiring the identification of impedance parameters, the Oustaloup approach offers the best compromise between the goodness of the obtained frequency response and the accuracy–complexity requirements.

Keywords: battery impedance model; equivalent circuit model; fractional-order model; time-domain implementation

Citation: Ospina Agudelo, B.; Zamboni, W.; Monmasson, E. A Comparison of Time-Domain Implementation Methods for Fractional-Order Battery Impedance Models. *Energies* **2021**, *14*, 4415. https://doi.org/10.3390/en14154415

Academic Editor: Hugo Morais

Received: 1 May 2021
Accepted: 18 July 2021
Published: 22 July 2021

1. Introduction

Energy storage systems (ESSs) are among the most critical components concerning the full adoption of renewable energy sources and electric transportation [1]. In such applications, energy management systems (EMSs) are required for coordinating the operations of systems with multiple energy generation resources, as often is the case with microgrids and hybrid ESSs, which employ multiple complementary energy storage subsystems. For the energy management of ESSs, the most commonly desired objectives are: improved energy efficiency, extended lifetimes for storage elements and compliance with constraints of internal energy storage modules, e.g., no over-charging and no over-discharging [2]. In the case of rechargeable batteries, those goals typically require accurate monitoring; state of charge (SoC) and state of health (SoH) estimations; and charging control, which can be accomplished by using dedicated battery management systems (BMS). The methods employed for these operations range from experiment-based ones, such as incremental capacity [3], to data-driven ones, including machine learning [4]. State estimation and control algorithms in a BMS may require battery models capable of approximating the battery's response under given operating conditions [5].

Equivalent circuit models (ECMs) are preferred over electrochemical or empirical models, as they approximate the dynamic behaviour of a battery with relatively high accuracy [6], while offering simplified descriptions of the complex physical and chemical processes occurring within batteries, by representing them with a set of lumped elements, including resistors, capacitors and inductors. ECMs have been widely used as parts of battery state estimation and charging control strategies. The adoption of ECM-based optimal state estimation, by directly solving the model-constrained optimisation problem [7] or using techniques such as the Kalman filter and all its variants [8], has enabled the implementation of online SoC and SoH estimations in EMS frameworks. The implementation of online optimal charging strategies, known for their high computational burden, has also been enabled by algorithms, including ECMs, by profiting from the accuracy–complexity trade off offered by such models [9].

The parameters of an ECM can be fitted from voltage and current data obtained during specific operating conditions [6]. However, these circuit components are often insufficient for modelling the dynamics of electrochemical processes such as charge and mass transfers and double layer capacitance in a battery, due to the spatial distribution of those processes [10]. This lack of physical significance may compromise the identification of relationships between health estimation and ECM parameters. Said drawback may be overcome by substituting capacitors in the ECM with constant phase elements (CPEs), defined in the frequency domain and analytically derived from the electrochemical principles of the diffusion processes. Such elements are often used to fit electrochemical impedance spectroscopy (EIS) data [11]. The reduced number of parameters in ECMs using CPEs instead of capacitors is mainly attractive for SoH estimation approaches based on the analysis of variations in the parameters associated with specific electrochemical processes [12].

The direct time-domain implementation of such models is particularly challenging, because they represent dynamic systems with non-integer-order derivative operations, hence the name fractional order models (FOM) [10]. The time-domain implementations of FOMs' electrical responses have been employed in applications such as online SoC estimation using fractional-order (FO) Kalman filters [13] and a time-domain characterisation of battery diffusion dynamics [14]. Three main implementation approaches have been identified: the multiple RC (mRC) circuit [15], the high-order integer transfer function [16] and the Grünwald–Letnikov (GL) fractional derivative [17] approximations. However, the literature is rather obscure about the motivation for the selection of one approach over the others. This work fills this gap by presenting a comparative study of the three approaches. The comparisons are made in terms of:

- *Accuracy:* by analysing the error of each implementation with respect to a reference analytical model for given input signals;
- *Complexity:* by writing the sets of equations required by each approach using similar structures and comparing the sizes of the arrays involved and the numbers of required additions and multiplications, in view of an embedded use of the FOMs;
- *Suitability for identification of impedance parameters* from time-domain measurements.

We used a simulation environment with a set of FOMs representing the middle-frequency range impedance of Li-ion batteries, normally represented by the so-called ZARC elements. Our study identifies which of the analysed approaches offers the best compromise between accuracy and computational burden for applications such as real-time simulations. The second goal was to understand for which of the three cases the time-domain identification of the FOM leads to a correct frequency-domain response, keeping the impedance model meaningful. This analysis may serve as a guide for the selection of implementation approaches for FOMs in EMS applications.

The paper is organised as follows. Section 2 introduces the battery impedance models based on constant phase elements, Section 3 presents, in detail, the three time-domain implementation approaches studied in this work. Then, the results of the comparative analysis in terms of accuracy and complexity are presented in Sections 4 and 5, respectively. Section 6 presents a time-domain identification study performed for the three implementation approaches. Finally, the conclusions are drawn in Section 7.

2. Battery Impedance Models

EIS analyses the impedance of battery cells in a specific range of frequencies, drawing conclusions about internal electrochemical processes with different time constants. A typical EIS test is performed by injecting a sinusoidal voltage or current signal; acquiring the cell response; and computing the impedance through the amplitude ratio and the phase difference between voltage and current signals. That experiment is repeated for several frequencies in a selected frequency range. Then, by calculating the impedance, a Nyquist plot, often called impedance plot, can be generated. If this is performed for multiple battery operating conditions (temperature, SoC and dc operating point), it is possible to visualise and quantify their influences over the frequency response. For a Li-ion battery cell, a typical impedance plot is shown in Figure 1 [18].

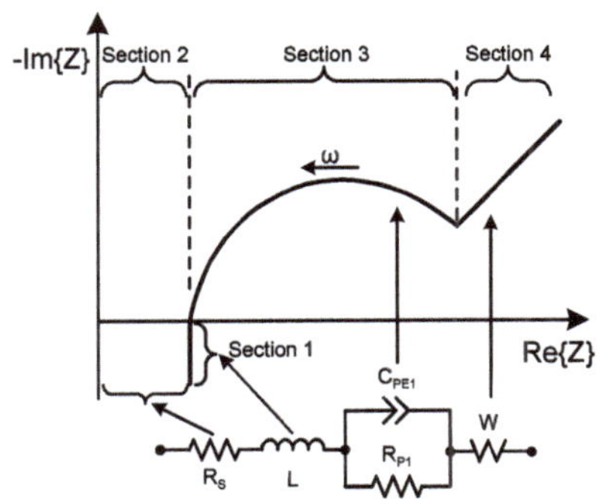

Figure 1. Qualitative impedance spectrum of a Li-ion cell and an ECM approximating it.

In the qualitative impedance plot shown in Figure 1, it is possible to identify four sections that can be associated with particular electrochemical processes [18]. In the first part, an inductive behaviour can be seen at high frequencies, related to the inductive reactances of metallic elements in the cell and wires. The presence of an ohmic resistance is revealed by the intersection with the real axis at a nonzero value. This corresponds to the sum of the current collectors, active material, electrolyte and separator resistances. The semi-circle-like section represents the double layer capacitance and charge transfer processes at the electrodes. Finally, at low frequencies, the main effect corresponds to the diffusion processes of the active material of the electrodes, which manifests as a section with a constant slope in the impedance plot. It is worth noting that measured spectra often show variations with respect to the qualitative curve presented. For example, the number of semi-circles can be increased to two or more, or the inductive part can exhibit a slope with an increasing real part. Moreover, the semi-circle section may tend to present a depression at its mid-point (non-constant radius) [19].

For a relatively accurate and meaningful reproduction of the impedance spectrum of a battery cell, the ECMs presented in Figure 1 can be considered [20]. The behaviour of the cell at high frequencies is represented by an ideal inductor L and the resistor R_s represents the ohmic resistance of the cell elements. The semi-circle section can be represented using a ZARC element, which corresponds to a parallel connection between a resistor and a CPE. The impedance of a CPE, Z_{CPE}, is presented in:

$$Z_{CPE}(\omega) = \frac{1}{Q(j\omega)^{\phi}}, \tag{1}$$

where Q corresponds to a generalised capacitance, ϕ to the depression factor, ω to the angular frequency and j to the imaginary unity [21].

The low-frequency response is represented by a Warburg impedance, which models a semi-infinite linear diffusion process. The Warburg impedance, namely, Z_W, is characterised by a 45° phase, and therefore can be also represented by a CPE:

$$Z_W(\omega) = \frac{A_w}{\sqrt{\omega}}(1-j) = \frac{\sqrt{2}A_w}{\sqrt{j\omega}} = \frac{1}{Q_w(j\omega)^{\phi_w}}, \tag{2}$$

where A_w is the diffusion coefficient [21], and the equivalent generalised capacitance and depression factor for the Warburg element are given by $Q_w = 1/(\sqrt{2}A_w)$ and $\phi_w = 0.5$, respectively.

CPEs are used instead of capacitors because when they are connected in parallel with a resistor forming a ZARC element, ϕ represents the depression often observed in the semi-circles in the impedance plot of a Li-ion cell. The factor ϕ can take values between zero and one; when $\phi = 0$, the ZARC element represents only an ohmic resistor, and if $\phi = 1$, the response of an RC element is obtained. The ZARC impedance is represented by:

$$Z_{ZARC}(\omega) = \frac{R_{p1}}{1 + R_{p1}Q_1(j\omega)^{\phi_1}}, \tag{3}$$

where R_{p1}, Q_1 and ϕ_1 characterise the resistance and CPE parameters of the ZARC element. The overall impedance of the circuit presented in Figure 1 is then defined by:

$$Z_{bat}(\omega) = j\omega L + R_s + Z_{ZARC}(\omega) + Z_W(\omega). \tag{4}$$

For the sake of simplicity, the discussion that follows focuses on the computation of the voltage $v(t)$ of the ZARC element for a given current signal $i(t)$. The presented results can easily be extended to the Warburg element using similar principles, and from there to the whole battery's impedance.

3. Time-Domain Implementation of the ZARC Element Response

In the framework of battery modelling, three main approaches for approximating the time-domain responses of elements with FO transfer functions have been proposed.

1. *Approximation 1.* The FO impedance is approximated with a series of parallel RC branches [22–24]. In [22], the mRC approximation was used for the implementation of two battery diffusion impedance models. A comparison in terms of the accuracy of the two models is presented. Similarly in [23], a CPE was used for approximating the diffusion dynamics of Li-ion batteries. A comparison of the identification performances of multiple implementations of the CPE based on the mRC approach is presented. The authors of [24] used the mRC approach for the approximation of the response of a ZARC element, in the framework of time-domain-based identification of impedance models for batteries.
2. *Approximation 2.* The approximation of the FO transfer function with a high-degree integer-order system, the Oustaloup (OU) approach being the most widely adopted [10,14,25]. In [10] and [25], this implementation was used for FOM identification with time-domain measurements. In [14], the approximation was used for studying and identifying the diffusion dynamics of Li-ion batteries.
3. *Approximation 3.* The approximation of a derivative of FO with a specific definition that allows for implementing difference equations, mainly based on the GL definition of the FO derivative [13,26,27]. Another study [26] analysed the performance of a FOM, implemented using the GL definition of the FO derivative, for the approximation of the battery voltage, and compared the results with a typical single RC model. Additionally, a method for the time-domain identification of the FOM was provided to the readers. In [13], the implementation was used in conjunction with a FO Kalman filter for online SoC estimation. Similarly, in [27], a GL-based implementation was used with an extended Kalman filter scheme for online estimation of SoC and SoH.

Flowcharts for the three implementation approaches are given in Figure 2. In the three approaches, the time response of the FOM is approximated by a set of discrete equations, corresponding to discrete state-space representations for the mRC and OU approaches and to a difference equation for the GL approach. These discrete implementations were evaluated in our simulation environment in terms of accuracy and complexity, as presented in the following sections. The main differences between the approaches lie in the parametrisation processes. For the GL approach, the coefficients in the difference equation are computed directly from the impedance parameters using predefined mathematical relationships, which are introduced in Section 3.3. As illustrated in Figure 2, the other two approaches require an initial approximation of the FO transfer function with an integer-order system, for which a discrete state-space representation can be obtained. The main difference between the mRC and OU approaches is the structure of the integer-order transfer function and the method used for the approximation of the FO transfer function, which are introduced in Sections 3.1 and 3.2, respectively.

In this section we will focus on defining difference equations for the computation of the ZARC voltage $v(t)$ for each one of the mentioned approaches. The resulting expressions have the form given by:

$$v[k] = Cx[k] + Di[k]. \tag{5}$$

Equation (5) allows one to compute the ZARC element voltage in the discrete instant k as a function of the input value i and the vector x. The vector x may be the states' vector or a set of previous values of v, depending on the implementation approach. The values of vector C and scalar D are functions of the parameters of each approximation, as is discussed in the following subsections.

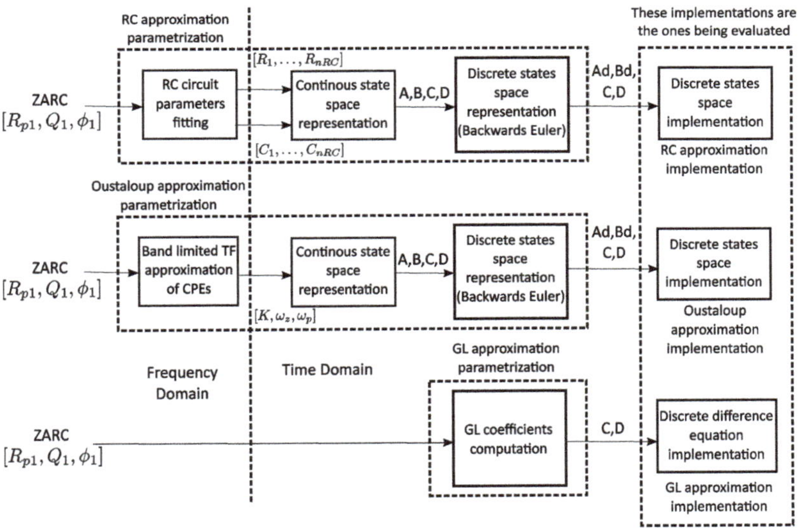

Figure 2. A summary of the FOM implementation approaches considered.

3.1. Approximation 1: Multiple RC Approach

For the approximation of the ZARC element impedance, series connections of multiple parallel RC circuits have been considered in the literature [15]. The values of the components for this kind of approximations are normally fitted from time measurements directly, by minimising the differences between experimental data and the model voltage output. However, for the case in which the initial point is an impedance model, the estimation process requires fitting the impedance spectra. The idea is to approximate the transfer function of the ZARC element with a set of RC parallel elements:

$$Z_{ZARC}(s) = \frac{R_{p1}}{1 + R_{p1} Q_1 s^{\phi_1}} \approx \sum_{h=1}^{n_{RC}} \frac{R_h}{1 + R_h C_h s} = Z_{RC}(s). \tag{6}$$

In (6), a transfer function representation is used for the impedance, using s as the Laplace complex variable. The approximation of the ZARC impedance $Z_{RC}(s)$ employs a set of n_{RC} RC branches. The parameters R_h and C_h represent the resistance and capacitance of the h-th RC parallel branch in the approximation of the ZARC impedance presented in (6), accounting for $2n_{RC}$ parameters to fit.

The values for the resistance and capacitance in (6) can be computed by minimising the difference between the ZARC impedance, given in (3), and the mRC approximation. One method for solving this minimisation problem is presented with detail in Appendix A. This method is employed from this point on during the parametrisation stage of the mRC-based implementations.

Once the values for R_h and C_h have been fitted, the continuous time response of the mRC circuit can be written in state-space representation:

$$\dot{x}_{RC}(t) = A_{RC}x_{RC}(t) + B_{RC}i(t) \tag{7}$$

$$v_{RC}(t) = C_{RC}x_{RC}(t) + D_{RC}i(t), \tag{8}$$

where the states' vector x_{RC} contains the RC elements voltages, $\dot{x}_{RC}$ represents the states' derivatives and v_{RC} is the approximation of the ZARC voltage using the mRC approach. Additionally, the matrices A_{RC}, B_{RC}, C_{RC} and D_{RC} are $n_{RC} \times n_{RC}$, $n_{RC} \times 1$, $1 \times n_{RC}$ and 1×1 matrices, respectively, given by:

$$A_{RC} = \mathrm{diag}\left(-\frac{1}{R_1C_1}, -\frac{1}{R_2C_2}, \dots, -\frac{1}{R_{n_{RC}}C_{n_{RC}}}\right) \tag{9}$$

$$B_{RC} = \begin{bmatrix} \frac{1}{C_1} & \frac{1}{C_2} & \cdots & \frac{1}{C_{n_{RC}}} \end{bmatrix}^T \tag{10}$$

$$C_{RC} = \begin{bmatrix} 1 & 1 & \cdots & 1 \end{bmatrix} \tag{11}$$

$$D_{RC} = 0. \tag{12}$$

For the discretisation of this set of equations, the derivatives are approximated using the backward Euler approximation for stability reasons [28]. The obtained set of difference equations can be rewritten and used for estimating the value of the ZARC element voltage in a discrete instant k, $v_{RC}[k]$, as a function of the current $i[k]$ and the RC elements voltages $x_{RC}[k]$:

$$x_{RC}[k] = (I_{n_{RC}} - TA_{RC})^{-1}x_{RC}[k-1] + T(I_{n_{RC}} - TA_{RC})^{-1}B_{RC}i[k] \tag{13}$$

$$v_{RC}[k] = C_{RC}x_{RC}[k] + D_{RC}i[k], \tag{14}$$

in which $I_{n_{RC}}$ is the $n_{RC} \times n_{RC}$ identity matrix, and T is the sampling period. Equations (13) and (14) can be used for implementing the response of the RC approximation given a current signal and a set of initial values for the RC branch voltages.

3.2. Approximation 2: Oustaloup Approach

This approach relies on approximating the transfer function of the CPE in the ZARC element using a transfer function with integer order n_{OU} (the same number of zeros and poles) in a given frequency range. One of the most popular methods for the approximation of the CPE transfer function with a rational transfer function of odd order n_{OU} was presented by Oustaloup et al. [16]. The approximation is valid in the frequency range $\omega \in [\omega_l, \omega_h]$, where ω_l and ω_h are the lower and higher frequency limits, respectively. Following this method, the transfer function of the CPE can be rewritten as:

$$Z_{CPE}(s) = \frac{1}{Q_1 s^{\phi_1}} = K_{CPE}\left(\frac{s}{\omega_c}\right)^{-\phi_1} \approx K_{CPE}\left(\frac{\omega_l}{\omega_c}\right)^{-\phi_1} \prod_{h=-N}^{N} \frac{1 + \frac{s}{\omega_{z(CPE)h}}}{1 + \frac{s}{\omega_{p(CPE)h}}} = Z_{OU(CPE)}(s),$$

where $Z_{OU(CPE)}(s)$ is the integer-order approximation of the CPE impedance using the OU approach; $\omega_c = \sqrt{\omega_l \omega_h}$ is the central frequency between the bounds of the range of interest; and K_{CPE} and N are given by $K_{CPE} = \frac{1}{Q_1 \omega_c^{\phi_1}}$ and $N = \frac{n_{OU}-1}{2}$. In order to estimate the n_{OU} zeros $\omega_{z(CPE)h}$ and n_{OU} poles $\omega_{p(CPE)h}$ in (15), Oustaloup et al. proposed [16]:

$$\omega_{z(CPE)h} = \omega_l \left(\frac{\omega_h}{\omega_l}\right)^{\frac{h+\frac{n_{OU}+\phi_1}{2}}{n_{OU}}}, \qquad \omega_{p(CPE)h} = \omega_l \left(\frac{\omega_h}{\omega_l}\right)^{\frac{h+\frac{n_{OU}-\phi_1}{2}}{n_{OU}}}. \tag{15}$$

The transfer function can be used to approximate the frequency response of the CPE. Then, the approximation $Z_{OU}(s)$ of the whole ZARC impedance using the OU approach is:

$$Z_{ZARC}(s) = \frac{R_{p1}}{1 + R_{p1}Q_1 s^{\phi_1}} \approx \frac{R_{p1}}{1 + \frac{R_{p1}}{Z_{OU(CPE)}(s)}} = Z_{OU}(s). \tag{16}$$

In order to obtain a time representation of the response of this approximation, the transfer function $Z_{OU}(s)$ needs to be rewritten in zero-pole-gain form, as expressed in:

$$Z_{OU}(s) = K_{ZARC} \prod_{h=1}^{n_{OU}} \frac{(s-\omega_{zh})}{(s-\omega_{ph})}, \tag{17}$$

where K_{ZARC}, ω_{zh} and ω_{ph} represent the gain, zeros and poles of the $Z_{OU}(s)$ transfer function, respectively. A time-domain state-space representation for the OU approximation with the structure presented in (17) was introduced in [28]:

$$\dot{x}_{OU}(t) = A_{OU}x_{OU}(t) + B_{OU}i(t) \tag{18}$$

$$v_{OU}(t) = C_{OU}x_{OU}(t) + D_{OU}i(t), \tag{19}$$

where x_{OU} and $\dot{x}_{OU}$ represent the system states and their derivatives, and v_{OU} is the approximation of the ZARC element voltage. The matrices A_{OU} $n_{OU} \times n_{OU}$, B_{OU} $n_{OU} \times 1$, C_{OU} $1 \times n_{OU}$ and D_{OU} 1×1, are given by:

$$A_{OU} = \begin{bmatrix} \omega_{p1} & 0 & 0 & \cdots & 0 \\ (\omega_{p2}-\omega_{z2}) & \omega_{p2} & 0 & \cdots & 0 \\ (\omega_{p3}-\omega_{z3}) & (\omega_{p3}-\omega_{z3}) & \omega_{p3} & \cdots & 0 \\ \vdots & \vdots & \vdots & \ddots & \vdots \\ (\omega_{pn_{OU}}-\omega_{zn_{OU}}) & (\omega_{pn_{OU}}-\omega_{zn_{OU}}) & \cdots & (\omega_{pn_{OU}}-\omega_{zn_{OU}}) & \omega_{pn_{OU}} \end{bmatrix} \tag{20}$$

$$B_{OU} = K_{ZARC}\left[(\omega_{p1}-\omega_{z1}) \quad (\omega_{p2}-\omega_{z2}) \quad \cdots \quad (\omega_{pn_{OU}}-\omega_{zn_{OU}})\right]^{\top} \tag{21}$$

$$C_{OU} = \begin{bmatrix} 1 & 1 & \cdots & 1 \end{bmatrix} \tag{22}$$

$$D_{OU} = K_{ZARC}. \tag{23}$$

Again, the backward Euler approximation was used to discretise the obtained state-space system. This discrete representation is obtained by replacing A_{RC}, B_{RC}, C_{RC} and D_{RC} with A_{OU}, B_{OU}, C_{OU} and D_{OU} in (13) and (14).

3.3. Approximation 3: Grünwald–Letnikov Approach

The ZARC element response can be approximated by adopting a FO derivative definition in the time domain. The FO differential equations representing the ZARC voltage can be obtained from transfer function (3) by replacing $Z_{ZARC}(s) = V(s)/I(s)$ and rewriting the equation as:

$$s^{\phi_1}V(s) = \frac{I(s)}{Q_1} - \frac{V(s)}{R_{p1}Q_1}, \tag{24}$$

where $V(s)$ and $I(s)$ represent the Laplace transform of the ZARC voltage and current, respectively. Then, by applying inverse Laplace transform, the FO differential equation is rewritten as:

$$\mathscr{D}^{\phi_1} v(t) = \frac{i(t)}{Q_1} - \frac{v(t)}{R_{p1} Q_1}. \tag{25}$$

$\mathscr{D}^{\phi_1}$ represents the derivative of FO ϕ_1. Among the multiple definitions of the FO derivative, the GL one is of particular interest, as it allows one to directly obtain difference equations for the approximations of the time response of a FO system [17]. The definition of the considered equation is:

$$\mathscr{D}^{\alpha} f(t) = \lim_{T \to \infty} \frac{1}{T^{\alpha}} \sum_{h=0}^{\lfloor \frac{t}{T} \rfloor} (-1)^h \binom{\alpha}{h} f(t - hT), \tag{26}$$

where the derivative of FO α of the causal function $f(t)$ is computed between 0 and t. In (26), T is the sampling period, $\lfloor \frac{t}{T} \rfloor$ represents the integer part of t/T and $\binom{\alpha}{h}$ represents the Newtonian binomial coefficients generalised to real numbers, computed as:

$$\binom{\alpha}{h} = \frac{\alpha(\alpha-1)(\alpha-2)\cdots(\alpha-h+1)}{h!} = \frac{\Gamma(\alpha+1)}{\Gamma(h+1)\Gamma(\alpha-h+1)}, \tag{27}$$

where $\Gamma(\cdot)$ stands for the gamma function, which works as a generalisation of the factorial operator for real numbers.

It is worth noting that, according to the GL definition in (26), the derivative of order α of the function at time t depends on all the values of that function in $[0, t]$, which is due to the non-local property of fractional derivatives [29].

By fixing the value of T to an appropriately low value for the application, and adopting the discrete variable k instead of the continuous time t, it is possible to obtain the first order discrete approximation of the FO derivative:

$$\mathscr{D}^{\alpha} f[k] = \frac{1}{T^{\alpha}} \sum_{h=0}^{k} (-1)^h \binom{\alpha}{h} f[k-h]. \tag{28}$$

This approximation may be used to obtain a difference equation for the numerical evaluation of function $f[k]$. The discrete version also requires all the data points of $f[k]$ since $k = 0$ for the computation of the derivative approximation, which may imply large memory requirements for simulations using this approach. This drawback may be addressed by applying the short-memory principle reported in reference [17], taking into account the behaviour of the signal in only the recent past, in the interval $[k - L, k]$, where L is the memory length. Applying this principle, we rewrite:

$$\mathscr{D}^{\alpha} f[k] = \frac{1}{T^{\alpha}} \sum_{h=0}^{L} (-1)^h \binom{\alpha}{h} f[k-h]. \tag{29}$$

This short-memory approximation allows one to implement numerical difference equations in cases in which the required memory is a critical constraint. Obviously, this introduces some inaccuracy, mostly manifested in the form of static error [30].

By replacing (29) in (25), we obtain the difference equation:

$$v_{GL}[k] = T^{\phi_1} \frac{R_{p1}}{R_{p1} Q_1 + T^{\phi_1}} i[k] - \frac{R_{p1} Q_1}{R_{p1} Q_1 + T^{\phi_1}} \sum_{h=1}^{L} (-1)^h \binom{\phi_1}{h} v_{GL}[k-h], \tag{30}$$

which allows one to compute the approximation of the ZARC element voltage using the GL approach v_{GL}. For the sake of comparison with the other implementation approaches, (30) can be written in matrix form:

$$v_{GL}[k] = C_{GL}x_{GL}[k] + D_{GL}i[k], \tag{31}$$

where $x_{GL}[k]$ is a $L \times 1$ vector with the L previous values of the ZARC voltage:

$$x_{GL} = \begin{bmatrix} v_{GL}[k-1] & v_{GL}[k-2] & \cdots & v_{GL}[k-L] \end{bmatrix}^{\top}. \tag{32}$$

The matrices C_{GL} and D_{GL}, having size $L \times 1$ and 1×1, respectively, are given by:

$$C_{GL} = -\frac{R_{p1}Q_1}{R_{p1}Q_1 + T^{\phi_1}} \begin{bmatrix} -\binom{\phi_1}{1} & \binom{\phi_1}{2} & \cdots & (-1)^h\binom{\phi_1}{h} & \cdots & (-1)^L\binom{\phi_1}{L} \end{bmatrix} \tag{33}$$

$$D_{GL} = T^{\phi_1}\frac{R_{p1}}{R_{p1}Q_1 + T^{\phi_1}}. \tag{34}$$

It is worth noting that for the matrix C_{GL}, the coefficients $(-1)^h\binom{\phi_1}{h}$ for the previous samples can be precomputed for the implementation of the ZARC voltage.

4. Accuracy Comparison

Ideally, the data required for the validation of approximations of the responses of FO battery impedance models must consider EIS and pulsed current tests both performed under similar SoC, SoH and temperature conditions. Due to the lack of availability of such data in the known datasets, and in order to perform accuracy comparisons between the analysed approximations, a reference model based on the analytical solution of FO differential equations was utilised for the generation of the reference data.

4.1. Reference Data Generation from the Analytical Solution of Fractional Differential Equations

Assuming that the current signal can be written as a set of steps:

$$i(t) = \sum_{h=1}^{N_u} U_h u(t - t_{uh}), \tag{35}$$

where $u(t)$ corresponds to the unit step function. Each one of the N_u current steps in $i(t)$ is characterised by an amplitude U_h and application time t_{uh}.

The ZARC element voltage can be expressed as [17]:

$$v(t) = \sum_{h=1}^{N_u} \frac{U_h}{Q_1}(t - t_{uh})^{\phi_1} E_{\phi_1,\phi_1+1}\left(-\frac{1}{R_{p1}Q_1}(t - t_{uh})^{\phi_1}\right)u(t - t_{uh}), \tag{36}$$

where the function $E_{\alpha,\beta}$ is the two-parameter Mittag–Leffler function, defined by a series expansion presented in Appendix B, which also includes a derivation of (36).

Equation (36) was implemented in Matlab® for the generation of the reference data. For the Mittag–Leffler function, the implementation introduced in [31] was employed. It is worth mentioning that such an analytical voltage representation is not suitable for online implementation, due to the limitations imposed by the assumed input signal and to the iterative nature of the Mittag–Leffler function computation, which makes the evaluation of a single data point highly demanding from a computational point of view.

4.2. Analysis of the Voltage Approximation Signals

The evaluation of the accuracy of the considered approximations requires one to perform an analysis of the differences between the responses of the reference model and the approximation of interest, for given ZARC element parameters and while using the same input current signal, as illustrated in Figure 3.

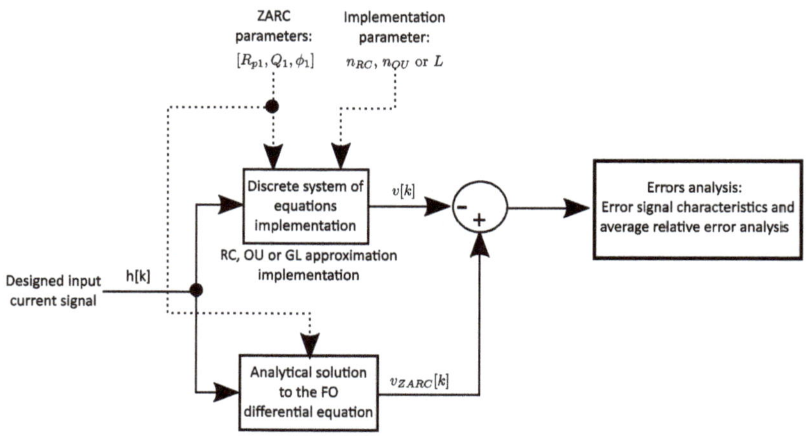

Figure 3. A schematic diagram representing the accuracy tests performed.

The test proposed in Figure 3 requires one to define an input current signal, a set of ZARC elements parameters and the order or memory length of the approximations. Tests such as this one were performed for six sets of ZARC element parameters, for currents generated with different sampling and dynamic characteristics and with variations of n_{RC}, n_{OU} and L. In order to generate the input currents and select the ZARC elements to be employed, first, we focused on the typical middle-frequency range of the dynamic response of Li-ion batteries, namely, the range between 0.01 and 200 Hz. This dynamic range is normally associated with the response of the double layer capacitance and the charge transfer resistance [18].

The considered input signals contained two stages: one aimed at evaluating the transient responses of the different approximations, and the second stage was for testing the steady state error. An input signal sample is presented in Figure 4. In this signal, the first stage has a total duration of 200 s, for which the amplitude and duration of each current pulse were selected randomly in the ranges $[-1, 1]$ A and $[0.5, 10]$ s, respectively. The second stage contains one single 0.5 A step with a duration of 500 s, with fixed 150 s rests before and after the step.

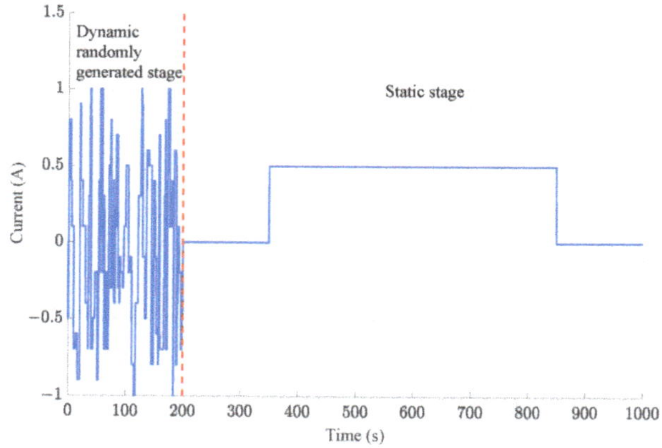

Figure 4. An example of the current profiles used during the accuracy tests.

The six sets of ZARC parameters we used are presented in Table 1. The ZARC parameters were selected for obtaining characteristic frequencies $\omega_0 = (1/(R_{p1}Q_1))^{1/\phi_1}$ covering the frequency range of interest, with $\omega_0 = 2\pi f_0$, while keeping the parameters' values inside typical ranges—namely, $R_{p1} \in [0.1, 100]$ mΩ, $Q_1 \in [1, 1000]$ F s$^{\phi_1 - 1}$ and $\phi_1 \in [0.5, 0.9]$ [32,33].

Table 1. Values of the parameters for the ZARC elements considered during the accuracy tests.

ZARC id	f_0 (Hz)	R_{p1} (mΩ)	Q_1 (F s$^{\phi_1-1}$)	ϕ_1
1	0.034	59.2	55	0.77
2	0.090	8.4	193	0.86
3	0.487	60.4	8	0.65
4	0.787	5.8	55	0.72
5	2.444	0.3	722	0.56
6	8.215	0.8	122	0.59

The evaluation of accuracy was performed for:

- Nine mRC-based approximations, employing n_{RC} values corresponding to the odd numbers between 3 and 19;
- Nine OU-based implementations, with odd orders n_{OU} between 3 and 19;
- Fourteen GL-based approximations, with L values between 5 and 10,000 samples (which correspond to time windows between 0.05 and 100 s).

Figures 5 and 6 present some examples of the voltage computed by mRC, OU and GL approximations for ZARC 4 employing $T = 0.01$ s. The plots in Figure 5a present the reference voltage and the voltages obtained for three mRC-based approximations during the first 20 s of the random stage of the accuracy test. In general, all the implementations were able to approximate the dynamics of the reference signal, with only appreciable differences for the approximation being of the lowest order. The error signals, presented in Figure 7a for this set of implementations, show spikes always under a few hundred µV during all the current steps, which reduce in magnitude as the order increases. Similar results can be observed for the OU-based implementations, as shown in Figure 5b and the first 200 s of the error signal in Figure 7b. The low-order OU approximations resulted in higher error magnitudes than the mRC ones. In the case of the GL approximations, as seen in Figure 5c, the lower L values caused higher offset errors. This is further illustrated by the error signal in Figure 7c during the first 200 s on which the spikes, at least for the lower memory lengths, seem to be wider than those in the mRC and OU approximations. An increase in L led to decreases in the magnitudes of the error signal, showing that adding terms to the sum in (30) leads to a better approximation of the analytical response, as expected.

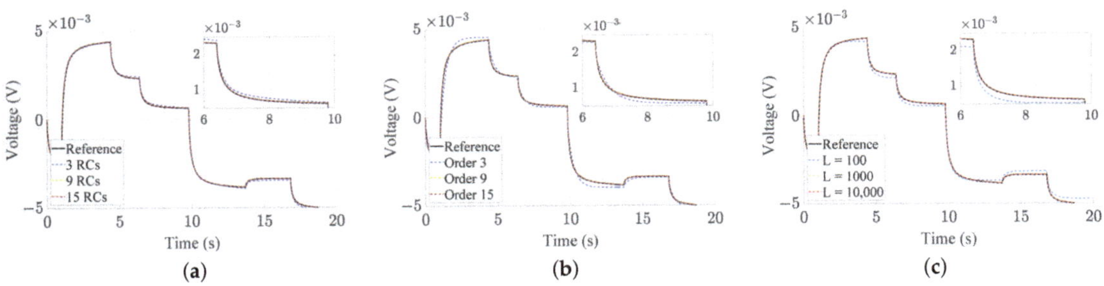

Figure 5. Examples of ZARC 4 voltages during the dynamic stage of the accuracy test. (**a**) mRC; (**b**) OU; (**c**) GL.

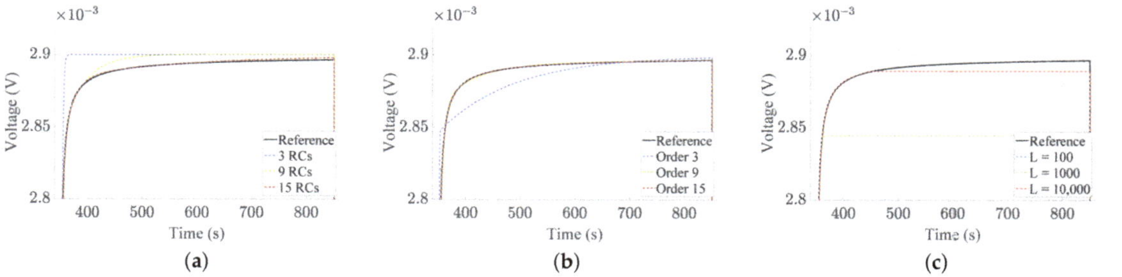

Figure 6. Examples of ZARC 4 voltages during the static stage of the accuracy test. (**a**) mRC; (**b**) OU; (**c**) GL.

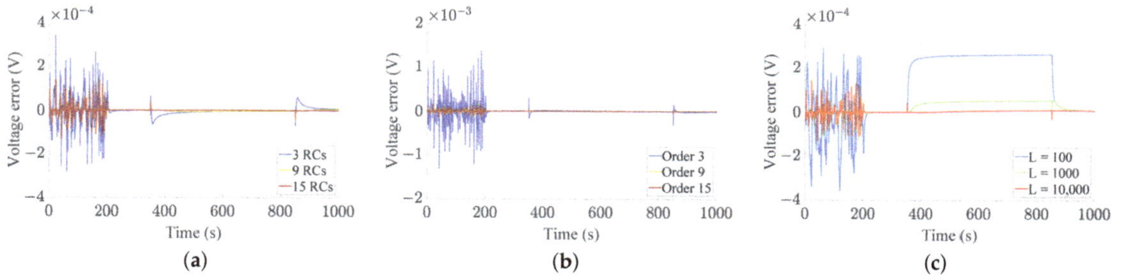

Figure 7. Examples of ZARC 4 voltage errors during the accuracy test. (**a**) mRC; (**b**) OU; (**c**) GL.

As is shown in the static state results presented in Figure 6a for three mRC-based implementations, increasing the order of the approximation reduced the errors during this stage. A higher value of n_{RC} allowed a better approximation of the distribution of time constants represented by the FO element, leading to an extension of the validity of the approximation over a wider frequency range. Similar considerations apply for the OU approximations, presented in Figure 6b, except for the worse performance at low orders, under nine, with respect to the mRC case. The offset error for the GL approximations is more evident during the static stage of the test, as shown in Figures 6c and 7c. The static error obtained for GL implementations with L under a few thousand for ZARC 4 is considerably higher than the errors obtained for the other approximations. This highlights the main drawback of the GL approximations using the short memory principle: by reducing the number of previous samples that are used for the computation, some level of inaccuracy appears, particularly in static state. For the sake of completeness, it is worth mentioning that Podlubny [17] proposed a relationship for estimating a suitable memory length for the approximation of the FO derivative presented in (28), given an expected error level.

4.3. *Effects of the Approximation Order and Memory Length on The Accuracy*

In order to analyse the effects of the parameters in each implementation approach, namely, n_{RC}, n_{OU} and L, on the approximation accuracy, the mean relative errors during the dynamic and the static stages were used as indicators. For ZARCs 1, 3 and 6, the mean relative errors during the dynamic and the static stages are presented in Figures 8a and 9a, respectively. The very similar results for the other ZARC elements are not reported for the sake of brevity.

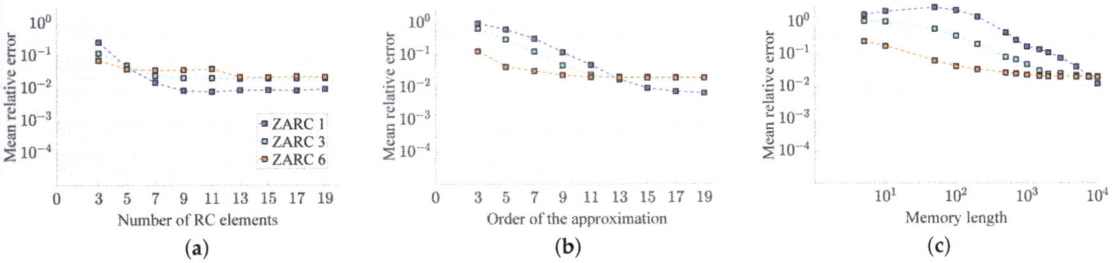

Figure 8. Average relative errors during the dynamic stage of the test. (**a**) mRC; (**b**) OU; (**c**) GL.

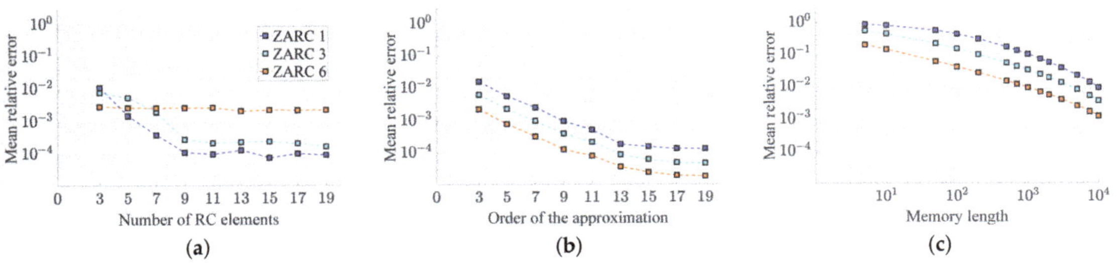

Figure 9. Average relative errors during the static stage of the test. (**a**) mRC; (**b**) OU; (**c**) GL.

The error during the static stage of the tests for the RC approximations remained under 1%, even for the approximations with fewer RC branches. This result was expected, as in the fitting procedure adopted in this work (Appendix A), the sum of the resistances in the mRC approximation was set to match the value of R_{p1}, leading to similar voltage drops in the response after the initial transitory. On the other hand, even if the values of the relative error during the dynamic stage tended to decrease with the number of employed RC elements, considerable improvements were only obtained up to $n_{RC} = 9$. After that point, further improvements could be achieved by decreasing the value of T. The observed behaviour at the highest orders may be associated with the variability introduced by the fitting procedure required for the computation of the parameters of the mRC case from the ZARC parameters. Nevertheless, for cases over $n_{RC} = 5$, the mean relative errors were always under 5%.

In the case of the OU-based approximations, the mean relative errors during the dynamic and the static stages are presented in Figures 8b and 9b, respectively. In general, for the OU approximations during the dynamic stage, orders n_{OU} higher than nine are required for reaching average relative errors under 5%. It is worth noting that, compared with the mRC approach, similar average relative errors were achievable in general, but with approximations of higher order. Again, the average relative errors for the static tests were almost always below 1%. For both stages of the accuracy tests, a monotonic reduction in the errors could be observed with increases in the approximation order, highlighting the advantage of computing the integer order approximation of the FO transfer function with a set of predefined equations instead of performing a fitting. This behaviour can be useful when trying to select the approximation order by analysing the accuracy–complexity trade off.

Then, for a set of GL-based approximations, with L values between 5 and 10,000 samples (which correspond to time windows between 0.05 s and 100 s), a similar accuracy analysis was performed. Figures 8c and 9c show the results obtained for the dynamic and static average relative errors obtained for this set of approximations. Regarding the results in the dynamic stage, only the higher memory lengths, over 500 samples, allowed us to reach values in the same order of magnitude as the ones obtained for the other approaches. In the case of the static stage, the average relative error is always about one order for magnitude higher, and it is evident that higher memory lengths or sampling times need to be considered for reducing the static error to a similar range. The errors were higher for the ZARC elements with slower dynamics, showing that slower systems require longer memory lengths to reach an acceptable accuracy level.

5. Computational Burden Comparison

A battery ECM can be used in an EMS as a part of the BMS state estimation structure or for battery simulation purposes during validation of energy management algorithms, particularly in real-time simulation scenarios. In both cases, considering that normally middle to low-end processing devices are often favoured due to budgetary restrictions, care needs to be taken in regard to the computational requirements of the battery model implementation. Here, we analyse those requirements in a general sense, by addressing the sizes of the matrices and the number of operations for each FOM implementation approach as indicators of required memory and computational complexity, respectively, in an eventual deployment.

For the three approaches, the time implementation relies on a set of matrix additions and multiplications. For mRC and OU-based approaches, the implementations consist of sets of equations in the form of (37) and (5) for the state and output equations, respectively. The discrete state equation established for the mRC approach can be generalised as:

$$x[k] = A_d x[k-1] + B_d i[k], \tag{37}$$

where A_d and B_d, namely, the discrete state and input matrices, are given by:

$$A_d = (I_{n_{RC}} - TA_{RC})^{-1} \tag{38}$$

$$B_d = T(I_{n_{RC}} - TA_{RC})^{-1} B_{RC}. \tag{39}$$

Conversely, the GL-based implementation relies only on a difference equation with the structure of (5), but in which $x[k]$ does not represent the system states vector but a vector with L previous values of the ZARC voltage.

Table 2 presents the sizes of the matrices and vectors used in the three approaches. Even if the mRC and OU approaches seem to be equivalents in terms of memory requirements, the simpler structure of the mRC can be exploited for the reduction of its memory requirements. Additionally, even if the number of arrays required for a GL implementation is lower and its size dependency given L seems simpler than those for the matrices in the other approaches, it is worth keeping in mind that in general for a given accuracy level L will take values in the range of hundreds or thousands, while n_{RC} or n_{OU} will be under 20.

Table 2. Sizes of the matrices in the state and output equations in the analysed implementations.

Element	mRC	OU	GL
A_d	$n_{RC} \times n_{RC}$ (Diagonal)	$n_{OU} \times n_{OU}$ (Lower triangular)	–
B_d	$n_{RC} \times 1$	$n_{OU} \times 1$	–
C	$1 \times n_{RC}$ (All-ones)	$1 \times n_{OU}$ (All-ones)	$L \times 1$
D	1×1 (zero)	1×1	1×1

Table 3 summarises the numbers of additions and multiplications required by each implementation approach. The specific structures of the matrices can be also exploited in the mRC and OU-based implementations, to refine the results presented in Table 3 by skipping the multiplications by zero and expressing the multiplication of column vectors by row vectors full of ones as an addition. The new operations count with this considerations is presented in Table 4.

Table 3. Numbers of additions and multiplications required for the analysed implementations.

Approach	Additions	Multiplications	Total
RC	$n_{RC}^2 + n_{RC}$	$n_{RC}^2 + 2n_{RC} + 1$	$2n_{RC}^2 + 3n_{RC} + 1$
OU	$n_{OU}^2 + n_{OU}$	$n_{OU}^2 + 2n_{OU} + 1$	$2n_{OU}^2 + 3n_{OU} + 1$
GL	L	$L + 1$	$2L + 1$

Table 4. Numbers of additions and multiplications with simplifications.

Approach	Additions	Multiplications	Total
RC	$2n_{RC} - 1$	$2n_{RC}$	$4n_{RC} - 1$
OU	$\frac{n_{OU}^2+3n_{OU}}{2}$	$\frac{n_{OU}^2+3n_{OU}+2}{2}$	$n_{OU}^2 + 3n_{OU} + 1$
GL	L	$L + 1$	$2L + 1$

Similarly to what was concluded for the array dimensions discussion, the expressions in Table 4 show that for the same order, a mRC implementation will require fewer operations than an OU one. It is worth mentioning that for the GL approach, even if there is not dependence on the square of L in the expressions for the number of operations required, the value of L needs to be considerably higher than the order for the other approaches to reach a given accuracy level.

The number of multiplications required for the evaluation of each type of implementation was used for assessing the computational burden in each case. Figure 10 plots the accuracy against the computational burden in terms of the number of multiplications for ZARC 4. The errors in static and dynamic stages are shown in Figure 10a,b, respectively.

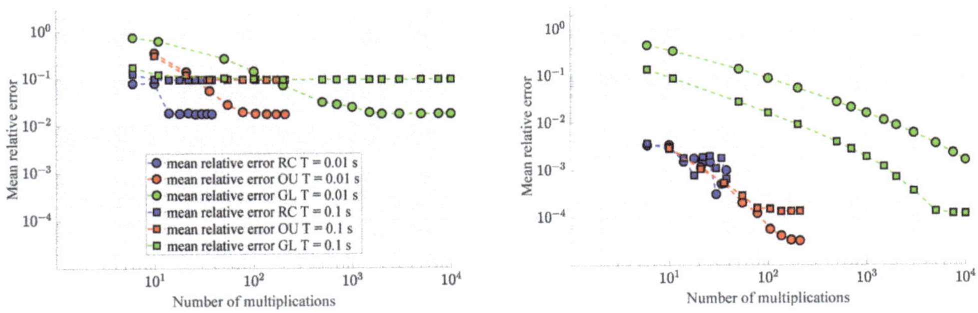

(a) (b)

Figure 10. Average relative errors vs. numbers of multiplications for mRC, OU and GL. (**a**) Dynamic stage; (**b**) static stage.

The curves for $T = 0.01$ s in Figure 10a show clearly how for a fixed mean relative error level in the dynamic stage of the accuracy tests, the number of required multiplications is always lower for the mRC approach, followed by the OU one. It is worth mentioning that the three approaches converge to values in the same order of magnitude for the mean relative error when increasing the complexity of the implementation; this may be an effect of the local truncation error due to the discretisation process. The asymptotic values of the analysed errors are comparable for a fixed sampling time. This can be ascribed to the fact that both the backward Euler and the GL derivative approximations are first-order approximations, leading to an $O(T)$ local truncation error, using big O notation [17]. To show the dependency on the sampling time of the identified error asymptotic values, the accuracy test was repeated using the current signal in Figure 4, but downsampled using $T = 0.1$ s. Those results are also summarised in Figure 10a, showing how the limit in the mean relative error is reached at a higher value, confirming the relationship between the sampling time and the maximum achievable accuracy.

The results presented in Figure 10b show that an increase in complexity has a more pronounced effect on the accuracy under static conditions, which tracks back to the requirement of higher orders or memory lengths for covering a wider time-constant range in the response approximation. In this case the previously observed oscillatory behaviour for higher orders in the mRC implementations is more evident for the two considered values of T. The effect of the local truncation error due to the discretisation process is also observed in this case, even if for lower values of the mean relative error.

From the results of the accuracy against complexity analysis, it can be clearly observed that the best accuracy–complexity compromise is offered by the mRC approach, allowing one to reach low error levels with a small computational burden. It is worth mentioning that this is the case provided good fitting of the ZARC element in the frequency domain can be performed. This is true for applications in which the model is used for battery simulation, but it is not the case when identification of impedance models from time measurements is required. In such instances, a good relationship between the parameters fitted from time-domain experiments with the frequency response is required. Thus, it is worth checking the suitability of the implementations for the time-domain identification of impedance parameters.

6. Analysis of the Suitability of Time-Domain Identification of the Battery Impedance

One of the main reasons for adopting battery FOMs is the capability of accurately approximating the voltage response while requiring a low number of parameters, which is of interest for tasks such as state estimation and battery characterisation. For this reason, in order to illustrate the applicability of the considered FOM implementation approaches in the framework of battery model identification using time-domain measurements, a set of time-domain FOM fitting tests were performed.

For the fitting tests, the reference voltage data $v[k]$ correspond to the random stage of the accuracy tests as the reference voltage signals presented in Figure 5. For all the ZARC elements and the implementations considered in the accuracy analysis, the associated parameters were fitted by minimising the mean square error between the reference voltage $v[k]$ and $v_x[k]$, which corresponded to $v_{RC}[k]$, $v_{OU}[k]$ or $v_{GL}[k]$ depending on the evaluated approximation. For all cases, the minimisation problem was solved in Matlab® using the default particle swarm optimisation (PSO) algorithm, implemented by the Matlab® function "particleswarm". The default PSO algorithm employs a number of particles automatically selected as the minimum between 10 times the number of parameters to be fitted and 100 particles; a function tolerance of 10^{-6}; and a maximum iteration number of 200 times the number of parameters to find [34].

The set of identified parameters changes depending on the approach evaluated. In the case of the GL-based implementations, the three parameters of the ZARC element, namely, R_{p1}, Q_1 and ϕ_1, can directly be identified due to the nature of this implementation, where the time response of the FO element is directly approximated, as presented in Section 3.3. This can be observed in the schematic of the identification procedure presented in Figure 11c, where the parametrisation process takes the identified values of the ZARC parameters as inputs for generating the vectors required for the time-domain implementation. Similarly, for the OU approach, the direct identification of the ZARC element parameters from time measurements is possible due to the direct relationship between R_{p1}, Q_1, ϕ_1 and the poles and zeros of the implemented integer-order transfer function, as introduced in Section 3.2. Figure 11b shows the implementation based on the OU approach, which was not modified for the fitting tests; only the source of the ZARC element parameters' changes, now being generated by the minimisation algorithm.

For the mRC approach, an initial frequency-domain fitting of the FO element impedance is required. For identification using time-domain measurements, this step cannot be performed. The resistance and capacitance values needed to be fit directly, as represented in the flowchart in Figure 11a.

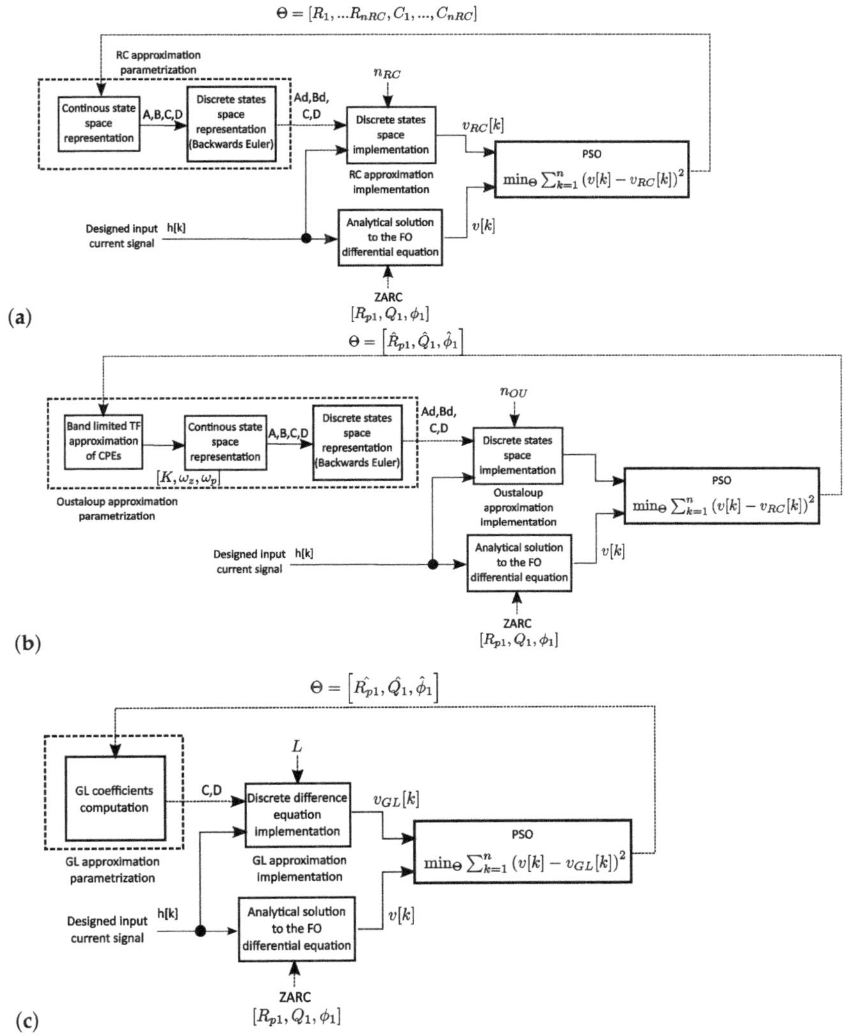

Figure 11. Block diagrams of the fitting procedures for the (**a**) multiple RC, (**b**) Oustaloup and (**c**) GL approximations.

On the one hand, for the OU and the GL approaches, the number of parameters to be identified was always 3, independently of the order n_{OU} or memory length L. On the other hand, the number of parameters to identify with the mRC approach increased with the order, being equal to $2n_{RC}$. This highlights the main drawback of the mRC approach: when fitting the time response of the FO element, overfitting issues may arise due to the high number of parameters.

For all the fitting tests, the mean of the relative error between the reference voltage and the response of the fitted approximation was computed as an indicator of the goodness of the time-domain fit. Additionally, the indicator of how close the obtained impedance is to the expected one in the range from 0.01 to 20 Hz is the following:

$$\delta_Z(\omega) = \frac{|Z_{ZARC}(\omega) - Z_{app}(\omega)|}{|Z_{ZARC}(\omega)|}. \tag{40}$$

Here, δ_Z corresponds to the relative distance between the impedance of the fitted approximation, called Z_{app}, and the one of the original ZARC element, Z_{ZARC}.

The results obtained for ZARC 3 are summarised in Figure 12 as a plot of the relative error in time against the one in frequency. Even if relative errors in time under 5% were obtained for multiple instances of each approach, similar results in frequency were only reached for OU and GL approximations. In the case of mRC-based approximations, a good fitting in time is not necessarily translated to a low distance between the approximation impedance and the original one. It is worth mentioning that the minimisation algorithm was not optimised, but despite this, good accordance between the original and approximated impedance was reached for the OU and the GL approximations and not for the mRC case. Similar results were reached for all the other ZARC elements, illustrating how for the mRC approach, the lack of a direct relationship between the approximation parameters—namely, the resistance and capacitance values—and the ZARC impedance parameters reduces the possibility of identifying an approximation in time domain that also has a frequency response close to the real one.

An on-board-oriented implementation of these identification methods should address additional issues. One of them is the noise affecting the current and voltage measurements. Such noise contributes to causing a bias in the identified parameters, which should be compensated for with advanced fitting algorithms, as shown, for instance, in reference [35]. The noise compensation is worthy to be a matter of future study.

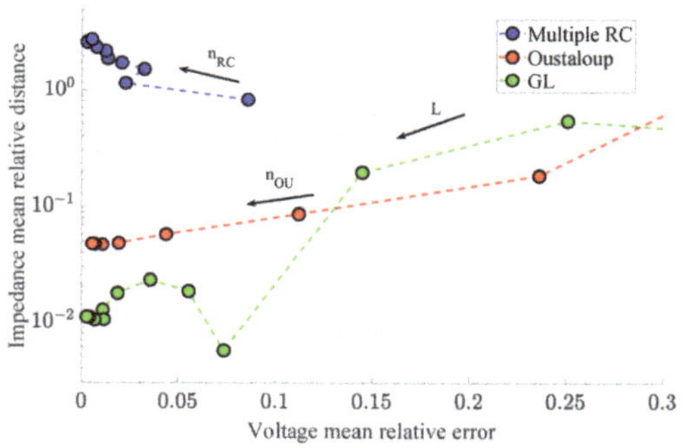

Figure 12. Fitting ZARC 3: average relative voltage error vs. average relative impedance distance.

7. Conclusions

In this work, the three main approaches adopted in the literature for the implementation of the time-domain response of battery FOMs were introduced and compared in terms of accuracy, computational requirements and suitability for the time-domain identification of battery impedance. The study was performed in a simulation framework with six different ZARC elements, which are normally used for the approximation of battery impedance in the middle-frequency range. The comparison was performed in a simulation environment where the reference solution was an analytical expression for the response of a ZARC element under a multiple-step current. The proposed expression, obtained using FO calculus theory, was used for generating the reference data required for the accuracy analysis of the considered implementations. Even if the discussion focused on ZARC elements, the results can be extended to the Warburg element and to the total battery impedance.

The primary results of the study can be summarised as follows.

- In terms of *accuracy under static conditions*, average relative errors under 0.1% were reached for all the evaluated ZARC elements using the three evaluated approaches. From the *computational complexity viewpoint*, these results were achieved with the mRC and OU approaches having similar computational requirements, whereas the GL approach often required a number of multiplications two orders of magnitude higher.
- In terms of *accuracy under dynamic conditions*, the mean relative errors converged to values in the same order of magnitude for the three approaches, when increasing the *complexity* of the implementation. The asymptotic errors were comparable for a fixed sampling time. For instance, the best mean relative error was around 2% using a sampling time of 0.01 s for all the methods.
- In terms of *suitability for identification from time-domain data*, all approaches well fit the time-domain voltage responses of the ZARC elements, with errors under 5%.
- However, in terms of *suitability to reproduce the frequency-domain impedance spectrum* from the parameters achieved in the time-domain identification, only OU and GL approximations reached errors of a few percent.
- The *best accuracy–complexity relationship* is offered by the mRC approach. It reached low error levels with the smallest computational burden. This should be the case as long as good fitting of the ZARC element in the frequency domain can be performed. This conclusion does not hold up if the starting point for the ZARC identification is time-domain measurements. In this latter case, the RC parameters may not lead to a correct frequency-domain response, and therefore, the best compromise for identification from time-domain data is represented by the OU approximation, which outperforms GL in terms of computational complexity.

The selection of the FOM implementation method depends on the application requirements. On the one hand, if the interest is only in the battery response simulation, the mRC approach offers the best accuracy–complexity compromise, which is desirable for real-time simulations oriented towards the validation of energy management algorithms. On the other hand, if the application requires accurate identification of the impedance parameters from time-domain measurements, the OU approach offers the best compromise among identifying the impedance model parameters, the complexity and the accuracy requirements.

Author Contributions: Conceptualization, B.O.A. and W.Z.; methodology, B.O.A. and W.Z.; software, B.O.A.; validation, B.O.A.; formal analysis, B.O.A.; investigation, B.O.A. and W.Z.; resources, W.Z.; data curation, B.O.A.; writing—original draft preparation, B.O.A. and W.Z.; writing—review and editing, B.O.A., W.Z. and E.M.; visualization, B.O.A.; supervision, B.O.A. and W.Z.; project administration, W.Z. and E.M.; funding acquisition, W.Z. and E.M. All authors have read and agreed to the published version of the manuscript.

Funding: This work was supported in part by funds from the the project "Holistic approach to EneRgy-efficient smart nanOGRIDS—HEROGRIDS" (PRIN 2017 2017WA5ZT3) within the Italian MUR 2017 PRIN programme; by the "Ex-WISCH e D-CODE" and FARB funds of the University of Salerno; and by funds of a public grant overseen by the French National Research Agency (ANR) as part of the "Investissements d'Avenir" program (reference: ANR-16-IDEX-0008).

Institutional Review Board Statement: Not applicable.

Informed Consent Statement: Not applicable.

Data Availability Statement: Not applicable.

Conflicts of Interest: The authors declare no conflict of interest.

Abbreviations

The following abbreviations are used in this manuscript:

BMS	Battery Management System
CPE	Constant phase element
ECM	Equivalent Circuit Model
EIS	Electrochemical impedance spectroscopy
EMS	Energy Management System
ESS	Energy Storage System
FO	Fractional Order
FOM	Fractional Order Model
GL	Grünwald–Letnikov
mRC	multiple RC
OU	Oustaloup
SoC	State of Charge
SoH	State of Health

Appendix A. Fitting of the Multiple RC Approaches' Parameters

The values of R_h and C_h, from the mRC approximation of the ZARC impedance introduced in Equation (6), can be computed by solving the minimisation problem:

$$\min_{\Theta_{RC}}\left\{ \left|Z_{ZARC}(j\omega) - Z_{RC}(j\omega)\right|^2 \right\}, \tag{A1}$$

where the array of parameters is given by $\Theta_{RC} = [R_1, \cdots, R_{M+1}, C_1, \cdots, C_{M+1}]$. This optimisation problem has been previously solved in the literature. The most widely accepted solution is the one proposed by [15], which presents the approximations of ZARC elements for given values of ϕ_1 using 1, 3 and 5 RC branches. That approximation relies on a number of optimisation parameters precomputed in [15] only for a few discrete values of ϕ_1. Here, we present a generalisation of this method for an odd number of RC branches $n_{RC} = M + 1$ and any positive value of $\phi_1 < 1$.

For the procedure, we take as starting point the parameters of the ZARC element, namely, R_{p1}, Q_1 and ϕ_1. We want to approximate the ZARC frequency response with $M + 1$ RC branches, with the values of the resistances and capacitors sorted as follows: $R_{RC} = [R_1, R_2, \cdots, R_{M/2+1}, \cdots, R_M, R_{M+1}]$ and $C_{RC} = [C_1, C_2, \cdots, C_{M/2+1}, \cdots, C_M, C_{M+1}]$. The generalisation can be obtained by defining the values of the resistance "in the middle", defined by the index $M/2 + 1$, according to:

$$R_{M/2+1} = \frac{K_1 R_{p1} \sin\frac{\phi_1\pi}{2}}{1 + \cos\frac{\phi_1\pi}{2}}. \tag{A2}$$

For all the remaining resistances, except those with indexes 1 and $M + 1$, and by taking advantage of the the symmetry of the ZARC element impedance spectra, the values are computed as fractions of the difference between R_{p1} and $R_{M/2+1}$, as expressed in:

$$R_h = R_{M+2-h} = \frac{K_h\left(R_{p1} - R_{M/2+1}\right)}{2}, \quad 2 \le h \le \frac{M}{2}. \tag{A3}$$

Finally the most "external" resistances are computed as:

$$R_1 = R_{M+1} = \frac{R_{p1} - \sum_{h=2}^{M} R_h}{2}. \tag{A4}$$

Then, the values of the capacitors can be estimated using the resistance and the characteristic frequency $\omega_0 = (1/(R_{p1}Q_1))^{1/\phi_1}$ values as follows:

$$C_{M/2+1} = \frac{1}{\omega_0 R_{M/2+1}} \tag{A5}$$

$$C_h = \frac{1}{K_{M/2+h}\omega_0 R_h}, \quad 1 \leq h \leq \frac{M}{2} \tag{A6}$$

$$C_h = \frac{1}{\frac{\omega_0}{K_{3M/2+2-h}} R_h}, \quad \frac{M}{2} + 1 \leq h \leq M + 1. \tag{A7}$$

While using this method for rewriting the optimisation problem presented in (A1), we can replace the parameters vector for $\Theta_{RC} = [K_1, \ldots, K_M]$, reducing the number of unknown parameters from $2M + 2$ (the number of resistors plus the number of capacitors) to M. The solution of the optimisation problem presented in (A1), and the set of Equations (A2)–(A7), were implemented in Matlab®, allowing us to obtain the parameters of the RC circuit that approximate the frequency response of an ideal ZARC element.

Appendix B. ZARC Element—Analytical Voltage Expression

For a ZARC element, the voltage $v(t) = \mathscr{L}^{-1}\{V(s)\}$ can be computed as [17]:

$$\begin{aligned} v(t) &= \mathscr{L}^{-1}\left\{ \frac{\frac{1}{Q_1}}{s^{\phi_1} + \frac{1}{R_{p1}Q_1}} I(s) \right\} = \frac{1}{Q_1}\left[t^{\phi_1 - 1} E_{\phi_1,\phi_1}\left(-\frac{1}{R_{p1}Q_1} t^{\phi_1} \right) \right] * i(t) \\ &= \frac{1}{Q_1}\int_0^t (t-\tau)^{\phi_1 - 1} E_{\phi_1,\phi_1}\left(-\frac{1}{R_{p1}Q_1}(t-\tau)^{\phi_1} \right) i(\tau) d\tau, \end{aligned} \tag{A8}$$

where $\mathscr{L}^{-1}\{X(s)\}$ is the inverse Laplace transform of $X(s)$, $E_{\alpha,\beta}$ is the two-parameter Mittag–Leffler function, the symbol $*$ stands for convolution and τ is the integration variable. The two-parameter Mittag–Leffler function is defined by the following series expansion:

$$E_{\alpha,\beta}(t) = \sum_{h=0}^{\infty} \frac{t^h}{\Gamma(\alpha h + \beta)}, \tag{A9}$$

and it can be seen as a generalisation of the exponential function e^t, which can be considered as a particular case of the $E_{\alpha,\beta}(t)$ function with $\alpha = \beta = 1$ [17].

Equation (A8) can be used for computing the voltage response of the ZARC element from the current signal. The main limitation of this solution is that it requires an expression for the current in order to be evaluated. Additionally, this analytical expression is very demanding computationally, as the computation for a given time value requires all the previous values. Simplifications of this analytical expression can be achieved by considering a step current signal $i(t) = Uu(t - t_u)$, where U and t_u are the step amplitude and time, and $u(t)$ represents the unit step function. By taking (A8), and using $I(s)$ for the Laplace transform of the current step, it is possible to obtain an expression that does not require the evaluation of an integral for the computation of the ZARC voltage:

$$\begin{aligned} v(t) &= \mathscr{L}^{-1}\{V(s)\} = \mathscr{L}^{-1}\left\{ \frac{\frac{1}{Q_1}}{s^{\phi_1} + \frac{1}{R_{p1}Q_1}} \frac{U}{s} e^{-t_u s} \right\} \\ &= \frac{U}{Q_1}(t - t_u)^{\phi_1} E_{\phi_1,\phi_1 + 1}\left(-\frac{1}{R_{p1}Q_1}(t - t_u)^{\phi_1} \right) u(t - t_u). \end{aligned} \tag{A10}$$

Using the superposition principle, it is possible to extend these results for current signals composed by combinations of sets of N_u step signals, meaning that they can be written as:

$$i(t) = \sum_{h=1}^{N_u} U_h u(t - t_{uh}), \tag{A11}$$

which is true for inputs such as square or pseudo-random binary sequence signals. If such an input current signal is considered, the resulting voltage signal for the ZARC element is given by:

$$v(t) = \sum_{h=1}^{N_u} \frac{U_h}{Q_1}(t - t_{uh})^{\phi_1} E_{\phi_1,\phi_1+1}\left(-\frac{1}{R_{p1}Q_1}(t - t_{uh})^{\phi_1}\right)u(t - t_{uh}). \tag{A12}$$

References

1. Mohammad, A.; Zamora, R.; Lie, T.T. Integration of electric vehicles in the distribution network: A review of PV based electric vehicle modelling. *Energies* **2020**, *13*, 4541. [CrossRef]
2. Tran, D.; Khambadkone, A.M. Energy management for lifetime extension of energy storage system in micro-grid applications. *IEEE Trans. Smart Grid* **2013**, *4*, 1289–1296. [CrossRef]
3. Ospina Agudelo, B.; Zamboni, W.; Monmasson, E. Application domain extension of incremental capacity-based battery SoH indicators. *Energy* **2021**, 121224. [CrossRef]
4. Severson, K.A.; Attia, P.M.; Jin, N.; Perkins, N.; Jiang, B.; Yang, Z.; Chen, M.H.; Aykol, M.; Herring, P.K.; Fraggedakis, D.; et al. Data-driven prediction of battery cycle life before capacity degradation. *Nat. Energy* **2019**, *4*, 383–391. [CrossRef]
5. Hidalgo-Reyes, J.I.; Gómez-Aguilar, J.F.; Escobar-Jiménez, R.F.; Alvarado-Martínez, V.M.; López-López, M.G. Classical and fractional-order modeling of equivalent electrical circuits for supercapacitors and batteries, energy management strategies for hybrid systems and methods for the state of charge estimation: A state of the art review. *Microelectron. J.* **2019**, *85*, 109–128. [CrossRef]
6. Qin, D.; Li, J.; Wang, T.; Zhang, D. Modeling and Simulating a Battery for an Electric Vehicle Based on Modelica. *Automot. Innov.* **2019**, *2*, 169–177. [CrossRef]
7. Wei, Z.G.; Hu, J.; He, H.; Li, Y.; Xiong, B. Load Current and State of Charge Co-Estimation for Current Sensor-Free Lithium-ion Battery. *IEEE Trans. Power Electron.* **2021**, *36*, 10970–10975. [CrossRef]
8. Lai, X.; Yi, W.; Cui, Y.; Qin, C.; Han, X.; Sun, T.; Zhou, L.; Zheng, Y. Capacity estimation of lithium-ion cells by combining model-based and data-driven methods based on a sequential extended Kalman filter. *Energy* **2021**, *216*, 119233. [CrossRef]
9. Ouyang, Q.; Wang, Z.; Liu, K.; Xu, G.; Li, Y. Optimal Charging Control for Lithium-Ion Battery Packs: A Distributed Average Tracking Approach. *IEEE Trans. Ind. Inform.* **2020**, *16*, 3430–3438. [CrossRef]
10. Alavi, S.M.; Birkl, C.R.; Howey, D.A. Time-domain fitting of battery electrochemical impedance models. *J. Power Sources* **2015**, *288*, 345–352. [CrossRef]
11. Krewer, U.; Röder, F.; Harinath, E.; Braatz, R.D.; Bedürftig, B.; Findeisen, R. Review—Dynamic Models of Li-Ion Batteries for Diagnosis and Operation: A Review and Perspective. *J. Electrochem. Soc.* **2018**, *165*, A3656–A3673. [CrossRef]
12. Tian, J.P.; Xiong, R.; Shen, W.X.; Sun, F.C. Fractional order battery modelling methodologies for electric vehicle applications: Recent advances and perspectives. *Sci. China Technol. Sci.* **2020**, *63*, 2211–2230. [CrossRef]
13. Ma, Y.; Zhou, X.; Li, B.; Chen, H. Fractional modeling and SOC estimation of lithium-ion battery. *IEEE/CAA J. Autom. Sin.* **2016**, *3*, 281–287. [CrossRef]
14. Nasser Eddine, A.; Huard, B.; Gabano, J.D.; Poinot, T. Initialization of a fractional order identification algorithm applied for Lithium-ion battery modeling in time domain. *Commun. Nonlinear Sci. Numer. Simul.* **2018**, *59*, 375–386. [CrossRef]
15. Buller, S. Impedance-Based Simulation Models for Energy Storage Devices in Advanced Automotive Power Systems. Ph.D. Thesis, RWYH Aachen University, Aachen, Germany, 2003.
16. Oustaloup, A.; Levron, F.; Mathieu, B.; Nanot, F.M. Frequency-band complex noninteger differentiator: Characterization and synthesis. *IEEE Trans. Circuits Syst. I* **2000**, *47*, 25–39. [CrossRef]
17. Podlubny, I. *Fractional Differential Equations*; Acadamic Press: San Diego, CA, USA, 1999.
18. Andre, D.; Meiler, M.; Steiner, K.; Wimmer, C.; Soczka-Guth, T.; Sauer, D.U. Characterization of high-power lithium-ion batteries by electrochemical impedance spectroscopy. I. Experimental investigation. *J. Power Sources* **2011**, *196*, 5334–5341. [CrossRef]
19. Jossen, A. Fundamentals of battery dynamics. *J. Power Sources* **2006**, *154*, 530–538. [CrossRef]
20. Andre, D.; Meiler, M.; Steiner, K.; Walz, H.; Soczka-Guth, T.; Sauer, D.U. Characterization of high-power lithium-ion batteries by electrochemical impedance spectroscopy. II: Modelling. *J. Power Sources* **2011**, *196*, 5349–5356. [CrossRef]
21. Barsoukov, E.; Macdonald, J.R. *Impendance Spectroscopy: Theory, Experiment, and Applications*, 3rd ed.; John Wiley & Sons: Hoboken, NJ, USA, 2018.

22. Gagneur, L.; Driemeyer-Franco, A.L.; Forgez, C.; Friedrich, G. Modeling of the diffusion phenomenon in a lithium-ion cell using frequency or time domain identification. *Microelectron. Reliab.* **2013**, *53*, 784–796. [CrossRef]
23. Ren, H.; Zhao, Y.; Chen, S.; Yang, L. A comparative study of lumped equivalent circuit models of a lithium battery for state of charge prediction. *Int. J. Energy Res.* **2019**, *43*, 7306–7315. [CrossRef]
24. Schröer, P.; Khoshbakht, E.; Nemeth, T.; Kuipers, M.; Zappen, H.; Sauer, D.U. Adaptive modeling in the frequency and time domain of high-power lithium titanate oxide cells in battery management systems. *J. Energy Storage* **2020**, *32*. [CrossRef]
25. Zou, Y.; Li, S.E.; Shao, B.; Wang, B. State-space model with non-integer order derivatives for lithium-ion battery. *Appl. Energy* **2016**, *161*, 330–336. [CrossRef]
26. Wang, B.; Li, S.E.; Peng, H.; Liu, Z. Fractional-order modeling and parameter identification for lithium-ion batteries. *J. Power Sources* **2015**, *293*, 151–161. [CrossRef]
27. Hu, X.; Yuan, H.; Zou, C.; Li, Z.; Zhang, L. Co-Estimation of State of Charge and State of Health for Lithium-Ion Batteries Based on Fractional-Order Calculus. *IEEE Trans. Veh. Technol.* **2018**, *67*, 10319–10329. [CrossRef]
28. Baranowski, J.; Bauer, W.; Zagorowska, M.; Dziwinski, T.; Piatek, P. Time-domain Oustaloup approximation. In Proceedings of the 2015 20th International Conference on Methods and Models in Automation and Robotics, Miedzyzdroje, Poland, 24–27 August 2015; pp. 116–120. [CrossRef]
29. Pooseh, S.; Almeida, R.; Torres, D.F. Discrete direct methods in the fractional calculus of variations. *Comput. Math. Appl.* **2013**, *66*, 668–676. [CrossRef]
30. Maamri, N.; Trigeassou, J.C. A comparative analysis of two algorithms for the simulation of fractional differential equations. *Int. J. Dyn. Control.* **2019**, *8*, 302–311. [CrossRef]
31. Podlubny, I. MATLAB Central File Exchange: Mittag-Leffler Function. Available online: https://www.mathworks.com/matlabcentral/fileexchange/8738-mittag-leffler-function (accessed on 25 June 2021).
32. Wang, Y.; Li, M.; Chen, Z. Experimental study of fractional-order models for lithium-ion battery and ultra-capacitor: Modeling, system identification, and validation. *Appl. Energy* **2020**, *278*, 115736. [CrossRef]
33. Dai, H.; Jiang, B.; Wei, X. Impedance characterization and modeling of lithium-ion batteries considering the internal temperature gradient. *Energies* **2018**, *11*, 220. [CrossRef]
34. Matlab. Particleswarm Documentation. Available online: https://it.mathworks.com/help/gads/particle-swarm-optimization-algorithm.html (accessed on 25 June 2021).
35. Wei, Z.; Dong, G.; Zhang, X.; Pou, J.; Quan, Z.; He, H. Noise-Immune Model Identification and State-of-Charge Estimation for Lithium-Ion Battery Using Bilinear Parameterization. *IEEE Trans. Ind. Electron.* **2021**, *68*, 312–323. [CrossRef]

energies

Article

Model Predictive Control for Efficient Management of Energy Resources in Smart Buildings

Francesco Simmini [1,*], Tommaso Caldognetto [1,2], Mattia Bruschetta [3], Enrico Mion [3] and Ruggero Carli [1,3]

1 Interdepartmental Centre Giorgio Levi Cases, University of Padova, Via Francesco Marzolo 9, 35131 Padova, Italy; tommaso.caldognetto@unipd.it (T.C.); carlirug@dei.unipd.it (R.C.)
2 Department of Management and Engineering, University of Padova, Stradella S. Nicola 3, 36100 Vicenza, Italy
3 Department of Information Engineering, University of Padova, Via Giovanni Gradenigo 6/B, 35131 Padova, Italy; mattia.bruschetta@dei.unipd.it (M.B.); enrico.mion.1@phd.unipd.it (E.M.)
* Correspondence: francesco.simmini@unipd.it

Abstract: Efficient management of energy resources is crucial in smart buildings. In this work, model predictive control (MPC) is used to minimize the economic costs of prosumers equipped with production units, energy storage systems, and electric vehicles. To this purpose, the predictive control manages the available energy resources by exploiting future information about energy prices, absorption and production power profiles, and electric vehicle (EV) usage, such as times of departure and arrival and predicted energy consumption. The predictive control is compared with a rule-based technique, herein referred to as a heuristic approach, that acts in an instant-by-instant fashion without considering any future information. The reported results show that the studied predictive approach allows one to achieve charging profiles that adapt to variable operating conditions, aiming at optimal performances in terms of economic cost minimization in time-varying price scenarios, reduction of rms current stresses, and recharging capability of EV batteries. Specifically, unlike the heuristic method, the MPC approach is proven to be capable of efficiently managing the available energy resources to ensure a full recharge of the EV battery during nighttime while always respecting all system constraints. In addition, the proposed control is shown to be capable of keeping the peak power absorption from the grid constrained within set limits, which is a valuable feature in scenarios with widespread adoption of EVs in order to limit the stress on the electrical system.

Keywords: efficient management; energy resources; heuristic approach; model predictive control; nanogrid; smart buildings

Citation: Simmini, F.; Caldognetto, T.; Bruschetta, M.; Mion, E.; Carli, R. Model Predictive Control for Efficient Management of Energy Resources in Smart Buildings. *Energies* **2021**, *14*, 5592. https://doi.org/10.3390/en14185592

Academic Editor: Maria Carmela Di Piazza

Received: 29 July 2021
Accepted: 1 September 2021
Published: 7 September 2021

1. Introduction

The deregulation of the electric power industry has recently become a topic of attention for investors, regulators, and other participants who aim to achieve decarbonization in the energy sector and a more efficient use of energy [1]. In this context, smartgrids allow the enhancement of the efficiency of electricity utilization from the points of generation to the end users, and they enable the participation of prosumers on the demand side. Home nanogrids usually consist of renewable energy sources (RESs) and energy storage systems (ESSs) that can be used to store or release power when needed. Electric vehicles (EVs) using electricity produced by renewable sources appear as a promising solution for a sustainable transportation sector in the near future [2–6]. EVs use rechargeable battery packs to store the energy needed for propulsion. Smart buildings can be equipped with charging points where EV batteries can be recharged, for example, during nighttime [7]. On the other hand, concerns have been raised that are related to the peak power required to allow a proper recharge of EV batteries [4].

In this field, demand response (DR) will play an important role in the coordination of energy production and consumption by prosumers [8,9]. DR programs can be categorized as (i) incentive-based, where measures affect prosumers' behavior by providing incentives, or (ii) price-based, where electricity price variations are used to induce the prosumers to correspondingly adapt their electricity usage [10–12]. Under this latter category, time of use (TOU) and real-time pricing (RTP) are the most commonly adopted approaches to retail pricing [9,13]. With TOU, electricity prices are changed in order to follow the shape of the demand (e.g., higher prices during peak-load periods), while with RTP, prices are modified to follow the trends of the electricity market.

In such a scenario, the flexible and efficient management of prosumers' energy resources is crucial in order to make DR a win–win solution for both the electrical system and the prosumers. The implementation of an efficient management system can be achieved through the development of advanced control strategies that aim to increase the demand flexibility and to minimize the economic expenditures of prosumers [14]. In this context, model predictive control (MPC) [15,16] represents one of the most promising control methodologies for achieving an efficient management of users' resources. MPC is an advanced control strategy whose aim is to minimize an objective function over a prediction horizon while always satisfying a set of system constraints; due to its versatility, it has been largely adopted in applications in power systems [17–23]. For example, in [24], an MPC method was developed for a microgrid equipped with photovoltaic (PV) sources and ESSs while taking both the economic expenditure given by the power exchange with the upstream grid and the battery's wear cost into account in the objective function. In that work, the results of the MPC were studied with respect to the difference between the purchase and sale prices, but the electricity price was kept constant. Varying electricity prices were considered in [25], in which an energy management system based on MPC was developed and experimentally tested. The experimental infrastructure was located at the FlexElec Laboratory in the University of Nottingham, and it comprised an ESS and PV panels. The target of the MPC was to minimize the electricity bill. In the study, particular attention was given to the economic performance of the predictive control as different parameters were varied, such as the ESS capacity and the PV size. In [26], the economic MPC developed was compared with a simple control method that was implemented in a rule-based manner. The performances of the developed MPC for different price scenarios were investigated in the paper, and it was underlined that, when compared with simple control solutions, the MPC presented advantages with time-varying prices and that such advantages increased as the energy price variability rose.

In this work, the predictive control is used to efficiently manage a home nanogrid equipped with a PV source, an ESS, and a connection to the upstream grid, with the specific task of recharging the battery of an EV while fulfilling all of the operational constraints. A typical scenario that assumes that the EV battery is recharged overnight to take advantage of the lower prices commonly applied by retailers is considered. By exploiting predictions of absorption and generation profiles, energy prices, times of departure and arrival of the EV, and daily EV battery energy consumption, the predictive control manages the available energy resources with a twofold aim:

- Minimizing the economic costs of the system;
- Fully charging the EV battery during nighttime while respecting the maximum exchanged power constraints at the point of connection with the grid.

The MPC is compared with a simple control method called the heuristic approach that, unlike the MPC, manages the system in a rule-based fashion without taking any future information into account. The two control strategies are compared by considering two different perspectives. The first one regards the economic performances of the approaches in two different price scenarios—a flat TOU rate and a steep TOU rate. In this context, the MPC proves to be a better solution than the heuristic approach because it achieves lower economic costs with greater economic savings for the steep TOU rate, which is characterized by a high purchase-price variability. In addition, rms current stresses are reduced thanks to the MPC approach. The second perspective regards the functionality of the control methods: the MPC is capable of fully recharging the EV battery during nighttime in scenarios in which the heuristic method is not. This merit is due to the smart management allowed by the MPC approach, which is shown to be capable of fully exploiting the available resources while still satisfying all system constraints.

The organization of the paper is as follows. In Section 2, the system model of the nanogrid considered is reported. The heuristic approach is described in Section 3. Section 4 describes the proposed energy management system based on MPC. In Section 5, the MPC approach and the heuristic method are compared by considering the economic costs obtained in a seven-day simulation period, while Section 6 proposes a functional comparison between the two control approaches. The conclusions are reported in Section 7.

2. System Model

The smart building under consideration is provided with a nanogrid that connects the energy sources and loads, as shown in Figure 1. The nanogrid is composed of:

- Connection with the upstream grid with which the nanogrid exchanges power P_g. The maximum power that can be requested from the grid is denoted by $P_{g,max}$, which is assumed to be greater than 0 W.
- Photovoltaic (PV) sources, which generate the output power P_{pv}.
- Loads, which absorb power P_l.
- An energy storage system (ESS), which generates power P_{st} and has a capacity $E_{st,N}$, maximum discharging power $P_{st,max}$, and maximum charging power $P_{st,min}$. The capacity range of the ESS is given by the closed interval $[E_{st,min}, E_{st,max}]$.
- An electric vehicle battery (EVB), which generates power P_{ev} and has a capacity $E_{ev,N}$. In this paper, P_{ev} is assumed to be non-positive ($P_{ev} \leq 0$), i.e., the EVB can not provide power to the other system resources. The closed interval $[E_{ev,min}, E_{ev,max}]$ represents the capacity range of the EVB.

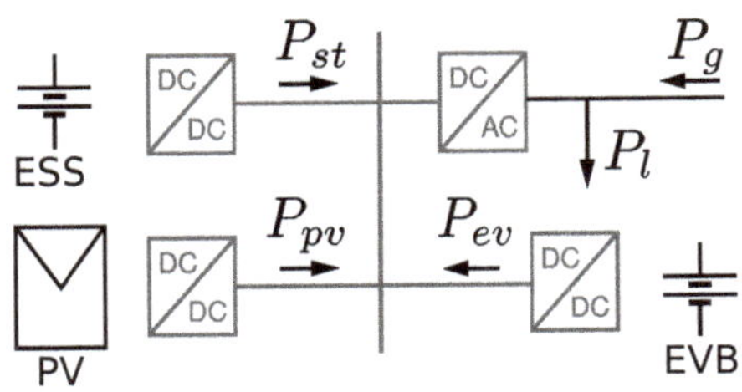

Figure 1. Smart building: power values under consideration.

Both the ESS and the EVB are modeled as a dynamic discrete-time system with ΔT as sampling interval:

$$E_{st}^{+} = E_{st} - \Delta T \, P_{st} \, , \qquad E_{ev}^{+} = E_{ev} - \Delta T \, P_{ev} \, , \tag{1}$$

where E_{st} and E_{ev} are the stored energy in the ESS and EVB, respectively, with superscript $^{+}$ referring to the value of the variable at the following time.

The model of the system with P_{st} and P_{ev} as inputs can be summarized in the following discrete-time equation:

$$\begin{bmatrix} E_{st}^{+} \\ E_{ev}^{+} \end{bmatrix} = \begin{bmatrix} 1 & 0 \\ 0 & 1 \end{bmatrix} \begin{bmatrix} E_{st} \\ E_{ev} \end{bmatrix} + \begin{bmatrix} -\Delta T & 0 \\ 0 & -\Delta T \end{bmatrix} \begin{bmatrix} P_{st} \\ P_{ev} \end{bmatrix} . \tag{2}$$

According to Figure 1, the power balance equation of the nanogrid is given in the following:

$$P_l = P_{pv} + P_{st} + P_{ev} + P_g \, . \tag{3}$$

3. Heuristic Approach

In this paper, the proposed MPC strategy is compared with a rule-based technique, which is called the heuristic approach hereinafter.

When the EV is connected to the nanogrid (i.e., during nighttime), the heuristic method devotes all the available resources to the charging process of the EVB; to this purpose, at each time step, given the current value of the battery's state of charge E_{ev}, by exploiting (1), the signal $\bar{P}_{ev}$ is computed as:

$$\bar{P}_{ev} = \frac{E_{ev} - E_{ev,max}}{\Delta T}\,. \tag{4}$$

When the EV is not connected to the nanogrid (i.e., during daytime), signal $\bar{P}_{ev}$ becomes:

$$\bar{P}_{ev} = 0\,\mathrm{W}\,. \tag{5}$$

The heuristic method manages the ESS to achieve $P_g = 0$, if possible; at each time step, it compares the values of P_{pv} and $P_l - \bar{P}_{ev}$:

- If $P_{pv} \leq P_l - \bar{P}_{ev}$, the shortage of power is satisfied by the ESS; when the ESS does not have enough power capacity or it is empty, the approach computes the grid power P_g to satisfy the excess demand. Then:
  - If P_g is lower than or equal to $P_{g,max}$, the EV charging power is given by:

    $$P_{ev} = \bar{P}_{ev}\,; \tag{6}$$

  - If P_g exceeds $P_{g,max}$, the grid power is set to satisfy the upper bound limit:

    $$P_g = P_{g,max}\,, \tag{7}$$

    and the resulting EV charging power is yielded by (3):

    $$P_{ev} = P_l - P_{pv} - P_{st} - P_g\,; \tag{8}$$

- If $P_{pv} > P_l - \bar{P}_{ev}$, the excess power is stored in the ESS; in case the ESS is full or does not have sufficient power capacity, the unused power is diverted into the grid. As far as the EVB is concerned, because, in this case, the following inequality holds:

  $$P_g \leq 0 < P_{g,max}\,, \tag{9}$$

  the actual power absorbed by the battery is given by:

  $$P_{ev} = \bar{P}_{ev}\,. \tag{10}$$

It is worth noting that, at each time step, the computed power $\bar{P}_{ev}$ is a fictitious signal; the actual EVB charging power P_{ev} is equal to $\bar{P}_{ev}$ if $P_g \leq P_{g,max}$.

The steps of the heuristic approach are shown in Figure 2.

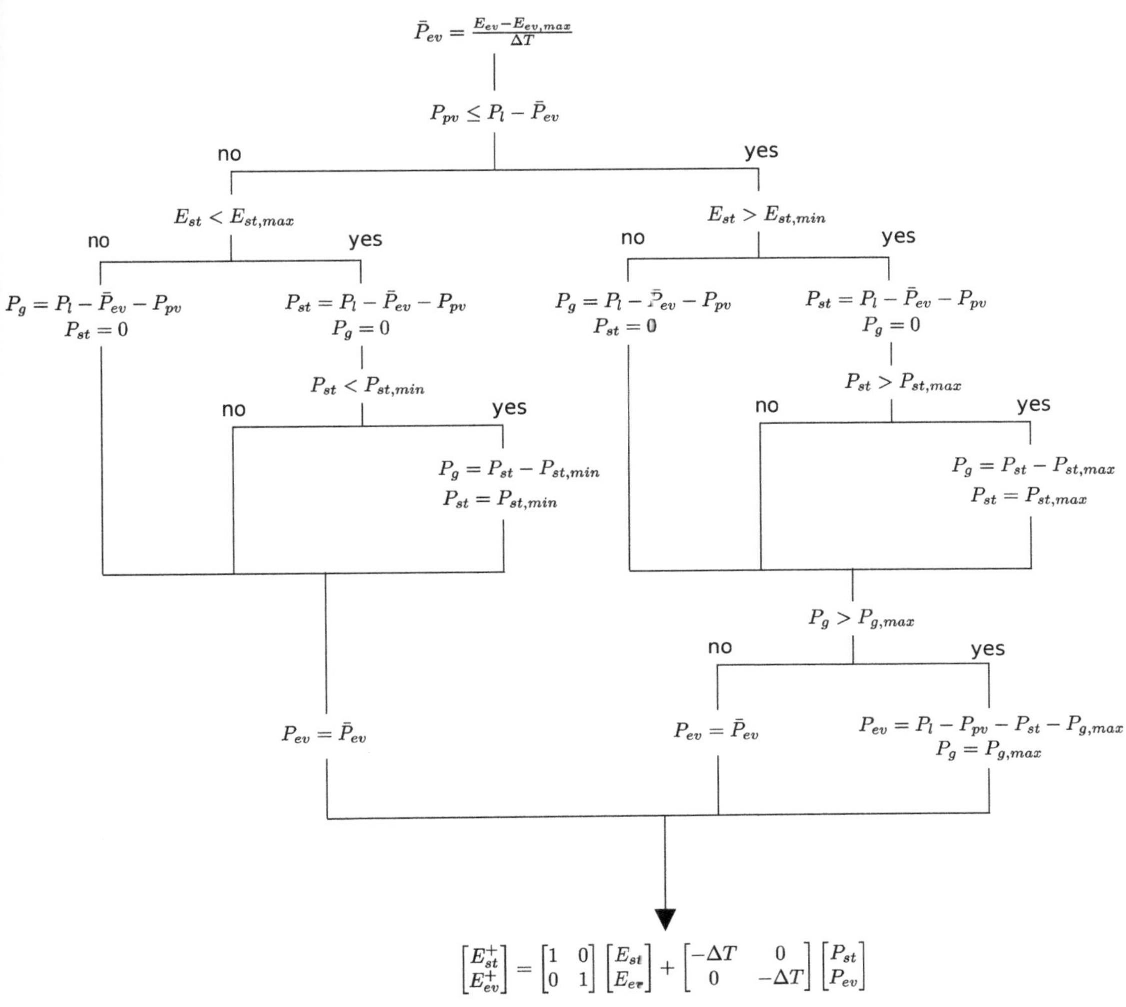

Figure 2. Heuristic approach: flowchart.

4. Model Predictive Control

Model predictive control is an advanced model-based control strategy whose aim is to optimize the evolution of a system over a future time span. At each sampling time, the predictive control computes the control signal by minimizing a cost function over a prediction horizon while taking system constraints into account. Only the first input of the signal is applied; then, the model state is updated and the optimization operation is iterated by adopting a receding horizon strategy.

In the proposed energy management system formulation, the MPC exploits future information about the expected power absorption and generation profiles, energy prices, times of departure and arrival of the EV, and daily EVB energy consumption with the aim of:

- Minimizing the overall economic costs of the system;
- Managing the system to achieve a full recharge of the EVB overnight.

In this paper, a particular case is considered in which future information about production and absorption profiles, departure and arrival times of the EV, and daily EVB energy consumption is assumed to be known, as the target is to highlight the advantages that can potentially be obtained by exploiting predictive approaches. Studying how prediction errors may degrade the performance of the MPC procedure will be the target of future work.

4.1. Cost Function

The MPC computes the control inputs that minimize the cost function J over a prediction horizon N_p:

$$\min_{\{P_{st}(k),P_{ev}(k)\}} J = J_e - J_f, \tag{11}$$

with

$$J_e = \Delta T \sum_{k=1}^{N_p} (c_p(k)\max(P_g(k),0) + c_s(k)\min(P_g(k),0) + c_{ESS}|P_{st}(k)|), \tag{12}$$

and

$$J_f = c_f E_{st}(N_p). \tag{13}$$

J_e is the economic cost over the prediction horizon and is given by the sum of the following three terms:

- The first term is the cost of purchasing energy; c_p is the purchase price, and is measured in EUR/kWh.
- The second term is the cost of selling energy (this term is negative, resulting in a positive revenue); c_s is the sale price, and is measured in EUR/kWh.
- The third term weighs the degradation of the ESS with the coefficient c_{ESS}, which is measured in EUR/kWh and will be defined in Section 4.3. This term considers the economic cost due to the wear of the storage unit.

The term J_f represents an economic value that is proportional to the final state of charge of the ESS in the prediction horizon via the positive coefficient c_f, which is measured in EUR/kWh. High values of c_f may lead to a fully charged ESS at the end of the prediction horizon, whilst low values lead to an empty ESS. The best setting of this coefficient depends on the price scenario under consideration and aims to mitigate the effect of a limited prediction horizon on the economic cost.

From (1), the following equation holds:

$$E_{st}(N_p) = E_{st}(0) - \Delta T \sum_{k=1}^{N_p} P_{st}(k). \tag{14}$$

Then, the cost function J can be written as follows:

$$J = \Delta T \sum_{k=1}^{N_p} \Big(c_p(k)\max(P_g(k),0) + c_s(k)\min(P_g(k),0) + c_{ESS}|P_{st}(k)| + c_f P_{st}(k)\Big). \tag{15}$$

4.2. System Constraints

The MPC minimizes the cost function in (15) in the prediction horizon under the following constraints for the ESS:

$$\begin{cases} E_{st,min} \leq E_{st} \leq E_{st,max} \\ P_{st,min} \leq P_{st} \leq P_{st,max} \end{cases}. \tag{16}$$

Considering the EVB constraints, when the EV is connected to the nanogrid (i.e., during nighttime), the following constraints hold:

$$\begin{cases} E_{ev,min} \leq E_{ev} \leq E_{ev,max} \\ E_{ev}(T_{d,i}) = E_{ev,max} \\ P_{ev} \leq 0 \end{cases}, \tag{17}$$

where $T_{d,i}$ refers to the departure time of the EV at the end of the i-th nighttime period; in this way, the MPC procedure ensures, if possible, a full EVB at the departure time. Otherwise, when the EV is not connected to the nanogrid (i.e., during daytime), the following equality constraint holds:

$$P_{ev} = 0\,. \tag{18}$$

As regards the grid power, the MPC optimizes the system under the following inequality constraint:

$$P_g \leq P_{g,max}\,. \tag{19}$$

4.3. Estimating the Economic Cost of ESS Wear

In the literature dealing with the economic optimization tasks for ESSs, a term in the cost function is usually added with the aim of preserving the life expectancy of the storage. In this paper, a common approach is adopted that consists of including a term that is proportional to the absolute value of the ESS's exchanged output power in the cost function [18]. The coefficient c_{ESS} in (15) is defined as the ratio between the ESS price and the total energy throughput bearable by the ESS in its whole usable life:

$$c_{ESS} = \frac{C_{ESS}}{2N_{cy}E_{st,N}}, \tag{20}$$

where C_{ESS} is the cost of the ESS, N_{cy} is its expected cycle lifetime, and $E_{st,N}$ is its capacity.

It is worth noting that since the aim of the MPC is to fully charge the EVB during nighttime with the constraint $P_{ev} \leq 0$, no EVB degradation cost is included in the cost function (15); the inclusion of a battery degradation cost would not modify the optimization results.

5. Economic Comparison

In this section, the economic performances of the MPC and the heuristic method are compared by considering different price scenarios. The first price scenario considers a TOU tariff with low purchase-price variations, and it is based on rates that are currently available in Italy. The second tariff considers typical Australian TOU rates with a high purchase-price variation between nighttime and daytime.

The nanogrid under consideration is provided with an ESS with a nominal capacity of 13.5 kWh and maximum output power of 4 kW, which corresponds to a commercial size for the considered application [27]; it is assumed that the ESS can be used in the range $[E_{st,min}, E_{st,max}]$, where $E_{st,min}$ and $E_{st,max}$ are equal to 2 and 13.5 kWh, respectively. The ESS is assumed to have an expected lifetime of 5000 cycles and a cost of 7030 EUR, which means that $c_{ESS} = 0.0521$ EUR/kWh.

The EV is provided with a battery with a nominal capacity of 42 kWh, which is an average and representative value considering the current storage capacities of EVs of small/medium size [28]. The EVB can be used in the range $[E_{ev,min}, E_{ev,max}]$, where $E_{ev,min}$ and $E_{ev,max}$ are equal to 6 and 42 kWh, respectively. A typical scenario is considered in which the driver leaves home in the morning and returns home in the afternoon or in the evening. The times of departure and arrival of the EV and the corresponding battery energy consumption during the week are reported in Table 1. An average consumption of 28.49 kWh per 100 statute miles is assumed, which is typical for the size of the considered EV, considering a driver traveling, on average, about 33.40 statute miles per day [29]. It is assumed that the EV connects to the nanogrid at the time of arrival.

Table 1. The departure and arrival times and EVB energy consumption considered.

Day of the Week	Departures	Arrivals	Energy Consumption
Wednesday	6 a.m.	18 p.m.	10.80 kWh
Thursday	6 a.m.	18 p.m.	7.20 kWh
Friday	6 a.m.	18 p.m.	10.80 kWh
Saturday	7 a.m.	21 p.m.	3.60 kWh
Sunday	7 a.m.	21 p.m.	19.80 kWh
Monday	6 a.m.	18 p.m.	10.80 kWh
Tuesday	6 a.m.	18 p.m.	3.60 kWh

In the simulations, the model sample time is 60 s, while the control sample time is set to one hour for both the MPC and the heuristic method. The prediction horizon is set to 24 h in the predictive controller; this choice represents a proper trade-off between the reliability of predictions and energy cost minimization; shorter prediction horizons would bring an increase in the economic costs.

The two control procedures are compared in a seven-day simulation period, where the load and PV data refer to an installation of a rated load power of 3 kW and a rated PV generation unit of 4 kW; the average daily energy absorbed by the loads is about 16.15 kWh/day, while the average daily PV energy generated is about 13.26 kWh/day.

5.1. Scenario 1: MPC—Flat Time-of-Use Rate

A TOU retail tariff typically applied in the Italian market is adopted in this scenario [30]; it is characterized by a slightly varying purchase price based on three time intervals: (i) high consumption (midday hours), (ii) low consumption (nighttime), and (iii) weekends. Prices in the second and third intervals usually coincide in the Italian market.

As regards the selling price, in Italy, a state-run net metering scheme for power injections of small generators is employed; excess PV energy is sold at market price, and then, at the end of the year, the prosumer receives a payment based on the injected and absorbed energy [31].

In this price scenario, the coefficient c_f is set to 0.17 EUR/kWh. It was verified that both lower and higher values can bring an increase in the economic cost; with lower values, the controller cannot exploit all of the available PV power to charge the ESS, while with higher values, the discharge of the ESS when demand exceeds production cannot be guaranteed, even in the case of the availability of energy in the ESS.

Figure 3 shows the energy prices and the predictive control simulation results. The cost of energy (Figure 3a) is 0.2642 EUR/kWh from 8 a.m. to 8 p.m. and 0.2424 EUR/kWh from 8 p.m. to 8 a.m. on working days, while on the weekend, it is a constant signal that is equal to 0.2424 EUR/kWh. The sale price changes on a daily basis, with small variations. In the simulations, it was assumed that the sale price for the upcoming day is known at 4 p.m.; in case of an unknown sale price at a specific time in the prediction horizon, the controller is fed with an average value based on the three previous days.

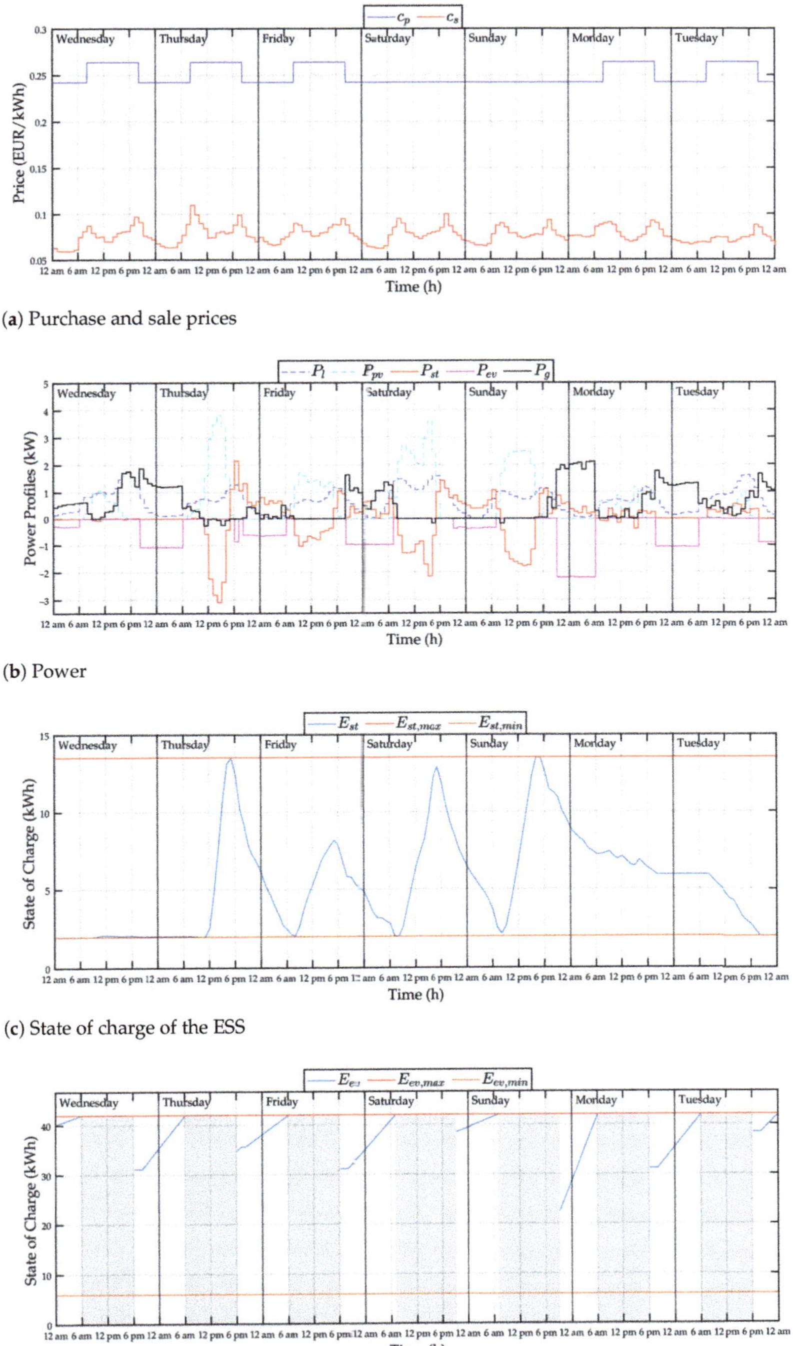

(**a**) Purchase and sale prices

(**b**) Power

(**c**) State of charge of the ESS

(**d**) State of charge of the EVB

Figure 3. MPC results with the flat TOU rate in a simulation of one week.

Power profiles are given in Figure 3b, while the states of charge of the ESS and the EVB are reported in Figure 3c,d, respectively (in Figure 3d, the time interval in which the EV is not connected to the nanogrid is shaded). It can be observed that that the ESS charges if the production exceeds the load during the daytime, while during the nighttime, power is driven from the ESS to the EVB, if possible (e.g., during the night between Saturday and Sunday); if the ESS cannot provide enough energy to charge the EVB, the predictive control purchases the required energy from the grid during the low-price periods. Notably, the MPC proves to be able to fully charge the EVB by smartly managing available energy resources with minimal economic expenditures.

5.2. Scenario 2: MPC—Steep Time-of-Use Rate

A tariff currently applied in the Australian electricity market is considered in this scenario [32]. It is characterized by a high variability in purchase price based on the following time intervals: (i) peak (midday hours), (ii) shoulders (morning, late evening, and weekend), and (iii) off-peak (nighttime). Furthermore, Australian retailers apply a daily charge between 0.90 and 1.20 AUD/day; in this case, a daily supply charge of 1.10 AUD/day is considered. As regards the selling price, in Australia, a flat rate that varies between 0.09 and 0.12 AUD/kWh is adopted; here, a sale price c_s equal to 0.12 AUD/kWh is applied. In the tests performed, an exchange rate of 0.6 EUR/AUD is considered.

In this price scenario, the coefficient c_f is set to 0.15 EUR/kWh. As described above, both lower and higher values can lead to an increase in the economic cost.

Figure 4 reports the simulation results for the MPC approach. On working days, the purchase price is 0.2376 EUR/kWh from 3 p.m. to 9 p.m. (peak), 0.1295 EUR/kWh from 7 a.m. to 3 p.m. and from 9 p.m. to 10 p.m. (shoulder), and 0.1086 EUR/kWh from 10 p.m. to 7 a.m. (off-peak). On the weekends, it is 0.1295 EUR/kWh from 7 a.m. to 10 p.m. (shoulder) and 0.1086 EUR/kWh from 10 p.m. to 7 a.m. (off-peak). The sale price is a constant signal that is equal to 0.072 EUR/kWh.

Differently from the previous price scenario, on Wednesday, the ESS charges during the off-peak time, thus exploiting the grid with the aim of satisfying the daytime demand during the peak price period, since the ESS is empty at the beginning of the considered period and the controller predicts a lack of generation in the prediction horizon; this is due to the daily purchase price difference, which is higher than $2c_{ESS}$ [26]. It is worth noting that the ESS is used only to satisfy the demand during the peak period of the working days, while on the shoulders and off-peak times, the grid is exploited to satisfy the load and to charge the EVB. Unlike in the previous scenario, there is no power transfer from the ESS to the EVB. This happens because, at the shoulder and off-peak times, the difference between the purchase and selling prices is lower than $2c_{ESS}$; the use of the ESS would increase the economic costs with respect to the solution of exploiting the grid to charge the EVB due to the cost of wear of the storage. Similarly to scenario 1, the MPC buys energy to charge the EVB during the lowest price period (from 10 p.m. to 7 a.m.).

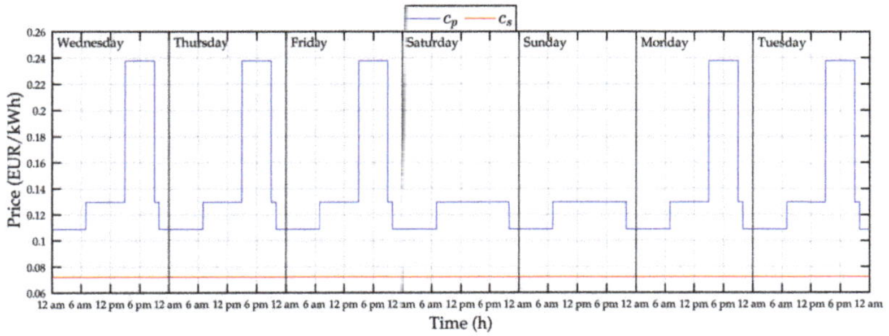

(a) Purchase and sale prices

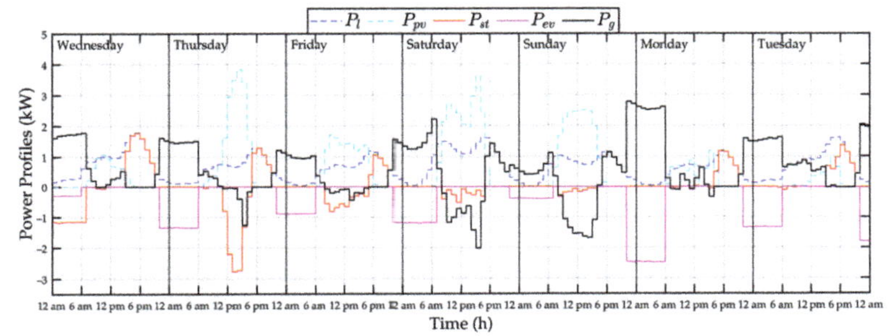

(b) Power

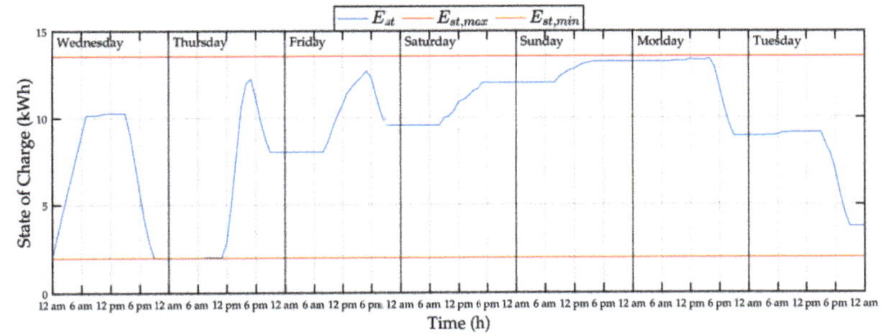

(c) State of charge of the ESS

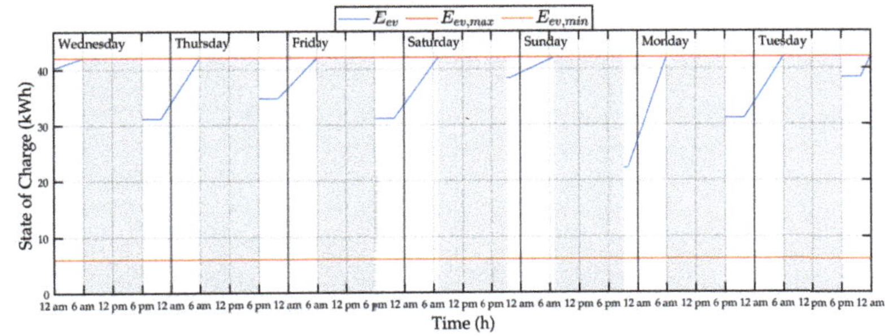

(d) State of charge of the EVB

Figure 4. MPC results with the steep TOU rate in a simulation of one week.

5.3. Simulation with the Heuristic Approach

Figure 5 reports the simulation results for the heuristic method. The same results are achieved for the two price scenarios considered, since the implementation of the algorithm of the heuristic approach does not depend on energy prices. During the daytime, when production exceeds absorption, excess PV power is diverted into the ESS or, if the storage is full, into the grid. Instead, when the PV power is lower than the load, the deficit is satisfied by the ESS or, if empty, by the grid. At the arrival time, the heuristic approach uses all of the available resources to charge the EVB, and then exploits the power provided by the ESS with an upper bound of 4 kW and that of the grid with an upper bound of 3 kW. As seen with the MPC, during the nighttime, the heuristic method is capable of fully recharging the EVB before the next departure time.

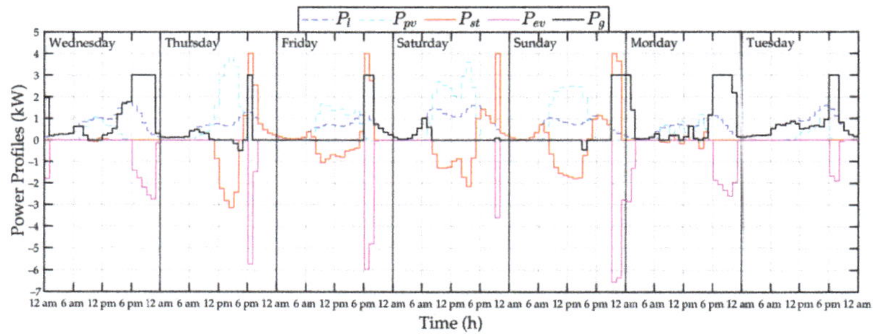

(**a**) Power

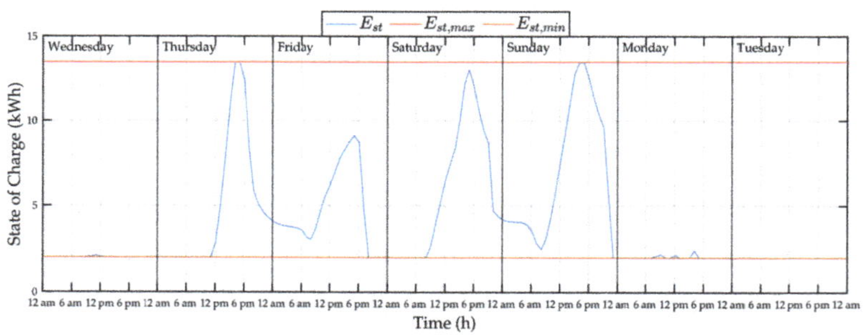

(**b**) State of charge of the ESS

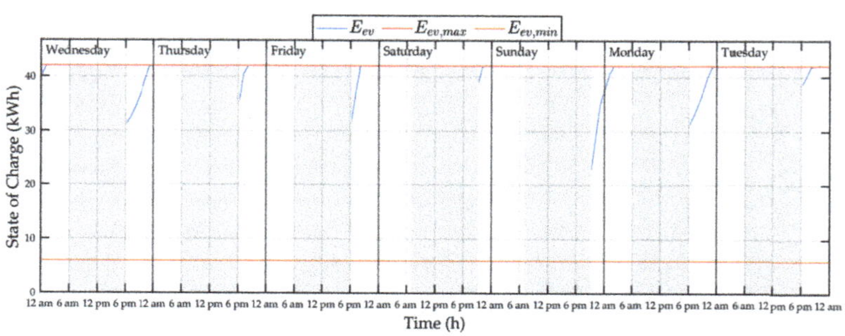

(**c**) State of charge of the EVB

Figure 5. Results of the heuristic approach in a simulation of one week.

5.4. Economic Costs

In this subsection, the performances of the predictive control and the heuristic method are evaluated by comparing the economic costs achieved in the simulations reported above.

The economic costs for the MPC and the heuristic approach in the considered price scenarios are reported in Table 2; the overall economic cost consists of the three terms defined in Section 4.1:

1. Energy purchase: cost of energy purchase,
2. Energy sale: cost of energy sale (this term is negative),
3. ESS wear: economic cost due to the wear of the ESS,

as well as the charge, which comes from a fixed daily value of 0.66 EUR/day in Scenario 2, while in Scenario 1, there is no charge.

Table 2. Economic costs in the considered interval of one week.

Control Methodology		Scenario 1: Flat TOU Rate					Scenario 2: Steep TOU Rate				
		Purchase	Sale	ESS	Charge	Overall	Purchase	Sale	ESS	Charge	Overall
MPC	(EUR)	22.13	−0.09	4.24	0	26.28	12.53	−1.65	2.74	4.62	18.24
Heuristic approach	(EUR)	22.69	−0.09	4.23	0	26.83	15.68	−0.08	4.23	4.62	24.45

As far as the overall economic costs are concerned, the economic savings of the MPC compared to the heuristic strategy are 2.05% in Scenario 1 and 25.40% in Scenario 2, rising up to 31.32% if the charge is not considered. The performance of the MPC increases with the increase in the daily price variability. In Scenario 1, which is characterized by a low daily variability of the purchase price, low savings are achieved, while in Scenario 2, much greater savings are achieved due to the high variation in the purchase price.

The MPC proves to be a better solution than the heuristic technique because it obtains lower economic costs by smartly managing available energy resources. Notably, such smart control actions are not possible with techniques that act in an instant manner, such as the heuristic approach considered here. It is worth remarking that the economic advantages obtained increase as the daily price variability rises [26].

Statistical results from the different cases are shown in Figure 6; the MPC substantially exhibits the advantage of providing lower I_g^2 (which is related to network losses) and lower I_{st}^2 and I_{ev}^2 (which are associated with battery degradation and converter losses) with respect to the heuristic method.

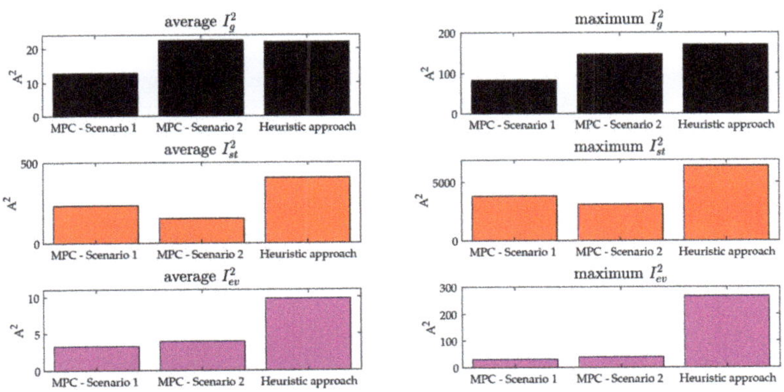

Figure 6. Statistical results in the considered interval of one week: $I_g = P_g/V_{g,N}$, $I_{st} = P_{st}/V_{st,N}$, $I_{ev} = P_{ev}/V_{ev,N}$, $V_{g,N}$ = 230 V, $V_{st,N}$ = 50 V, $V_{ev,N}$ = 400 V. $V_{g,N}$, $V_{st,N}$, and $V_{ev,N}$ refer to the nominal voltage of the grid, the ESS, and the EVB, respectively.

6. Functional Comparison

In this section, the MPC approach and the heuristic method are compared from a functional perspective. As in the example in the previous section, a seven-day simulation period is considered. The ESS and the EVB have the same characteristics as those introduced in Section 5. The planned schedule of departures and arrivals of the EV is provided in Table 3. Unlike in the example in the previous section, a much greater discharge of the EV battery in three days of the week (i.e., Friday, Saturday and Sunday) is considered. During the daytime, the EV is assumed to exploit all available energy stored in the battery (36 kWh).

Table 3. Functional comparison: planned departure and arrival times and EVB energy consumption.

Day of the Week	Departures	Arrivals	Energy Consumption
Wednesday	6 a.m.	18 p.m.	10.80 kWh
Thursday	6 a.m.	18 p.m.	7.2 kWh
Friday	6 a.m.	21 p.m.	36 kWh
Saturday	7 a.m.	22 p.m.	36 kWh
Sunday	7 a.m.	20 p.m.	36 kWh
Monday	6 a.m.	18 p.m.	10.80 kWh
Tuesday	6 a.m.	18 p.m.	3.60 kWh

In the tests performed here, the control sample time is set to one hour for both control approaches; the MPC is implemented with a one-day prediction horizon. As far as the grid power is concerned, $P_{g,max}$ is set to 3 kW.

In the simulations, the average daily energy absorbed by the loads is about 16.15 kWh/day, while the average daily energy generated by the PV panels is about 16.76 kWh/day.

Figure 7 shows the simulation results for the MPC approach in the flat TOU rate scenario. It is worth noting that, on Friday, Saturday, and Sunday, MPC successfully manages the energy resources (i) to fully charge the EVB during the nighttime and (ii) to respect the upper bound of 3 kW for the maximum power absorbed from the grid. On the one hand, during the daytime on Friday and Sunday, since there is lack of production, the predictive controller uses the grid to charge the ESS until reaching the state of charge that ensures a full EVB at the departure time of the following day. On the other hand, on Saturday, the MPC smartly manages available resources by exploiting the grid to satisfy the demand in the afternoon with the aim of leaving enough energy for nighttime charging in the ESS.

Figure 8 shows the simulation results for the heuristic approach. It is worth noting that, unlike the MPC, the heuristic approach is not able to ensure a full EVB before the departure times on Saturday, Sunday, and Monday. In detail, at the arrival time on Friday and Sunday, the ESS is empty, so it cannot provide energy to the EVB during the next night. On Saturday, the heuristic approach badly manages the ESS, since it exploits the stored energy in the afternoon before the arrival time of the EV; the remaining energy at the arrival time is not enough to guarantee a full EVB on Sunday morning. As a consequence of poor resource management, there is a decrease in the mileage with respect to what is planned (Table 3) on Saturday and Sunday because the driver has to satisfy the lower bound constraint of the EVB (6 kWh). Indeed, differently from the MPC case, the actual EVB energy consumption obtained with the heuristic method decreases by about 26% on Saturday and 9% on Sunday. It is worth remarking that, in this scenario, an economic comparison between the MPC and the heuristic method would be meaningless, as the mileage of the EV is different in the two cases.

In the above example, the heuristic approach shows the functional disadvantage of not being able to fully charge the EVB during the nighttime; on the other hand, this is possible through the use of predictive control techniques that manage the available resources by exploiting future information. The previous example highlights the potential for MPC in future contexts, since, by smartly managing available energy resources, it can guarantee the desired functionalities, which cannot be always ensured by simple heuristic techniques. Moreover, while satisfying the additional requirements of EV charging, the MPC is shown to be capable of keeping the peak power absorption from the grid constrained within nominal limits, which is an aspect of relevant concern considering the expected widespread use of electric vehicles.

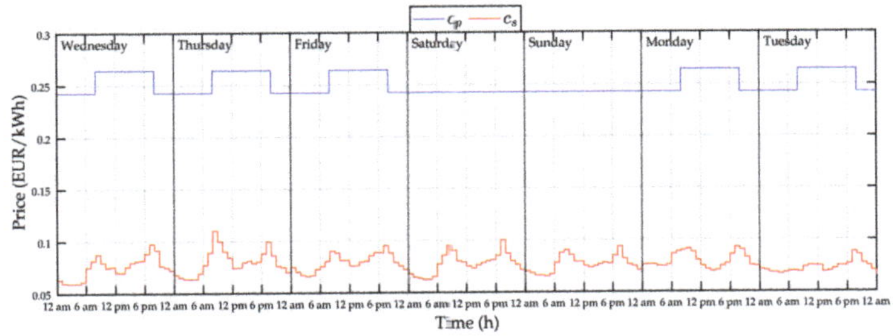

(**a**) Purchase and sale prices

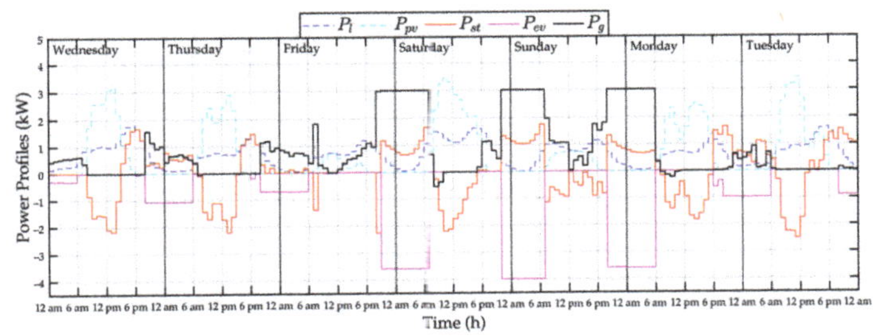

(**b**) Power

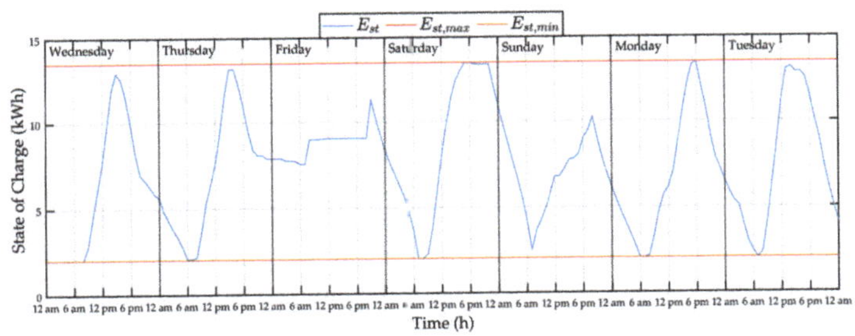

(**c**) State of charge of the ESS

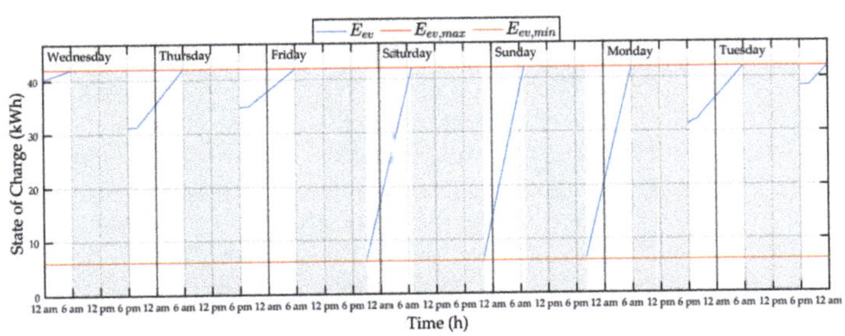

(**d**) State of charge of the EVB

Figure 7. Functional comparison: results of the MPC with the flat TOU rate in a simulation of one week.

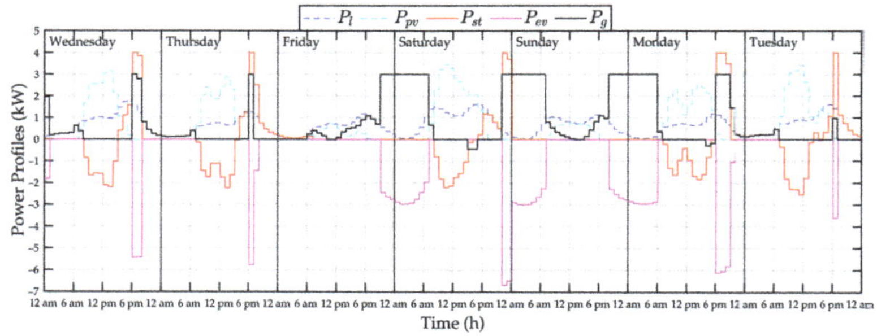

(a) Power

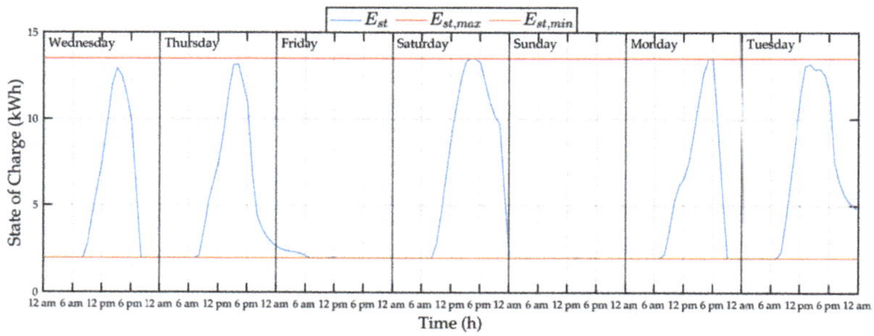

(b) State of charge of the ESS

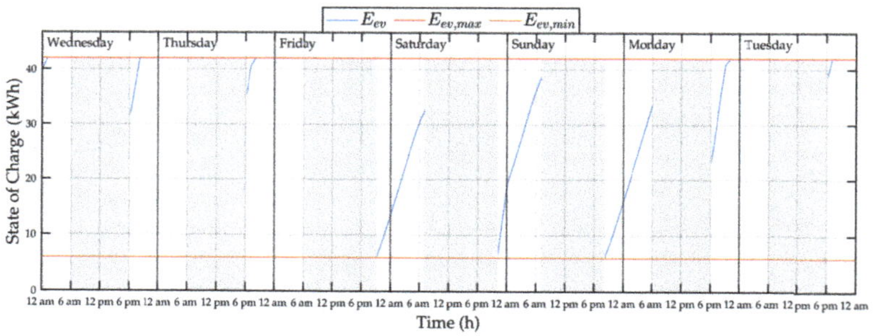

(c) State of charge of the EVB

Figure 8. Functional comparison: results of the heuristic approach in a simulation of one week.

7. Conclusions

This paper shows the application of model predictive control for the efficient management of prosumers' energy resources in the presence of local energy storage capabilities and an electric vehicle (EV). The aim of the predictive control developed here is the minimization of the economic cost with the constraint of ensuring a full EV battery at departure times. The proposed predictive approach is compared with a heuristic method that manages the available resources in an instantaneous rule-based manner. The simulation results show that, on the one hand, the MPC provides lower costs in scenarios in which both strategies are able to fully charge the EV battery during the night. On the other hand, it can guarantee a full recharge in scenarios in which the heuristic method is not capable of doing so. In summary, the simple scenario considered here allowed us to highlight the remarkable advantages of using predictive approaches in such an application, even while considering different energy pricing schemes. It is shown that the MPC algorithm, through optimal management of the available resources, allows one to (i) to meet functional objectives, such as the full recharge of the EV battery, (ii) allow the economic management of local resources, (iii) reduce rms current values at the interface with the mains and the local energy storage system, and (iv) keep the power exchanged with the mains within nominal ranges.

Future developments of the presented research may include the study of the effect of prediction errors on the performance of the MPC procedure and a sensitivity analysis of the MPC's performance as different parameters vary (e.g., length of the prediction horizon, energy storage capacities, PV system size).

Author Contributions: Conceptualization, F.S., T.C., M.B., E.M. and R.C.; methodology, F.S. and T.C.; software, F.S.; validation, F.S.; investigation, F.S. and T.C.; writing—original draft preparation, F.S.; writing—review and editing, F.S., T.C., M.B. and E.M.; supervision, R.C. All authors have read and agreed to the published version of the manuscript.

Funding: This research was funded by the Interdepartmental Centre Giorgio Levi Cases, University of Padova, NEBULE project.

Institutional Review Board Statement: Not applicable.

Informed Consent Statement: Not applicable.

Data Availability Statement: The data presented in this study are available on request from the corresponding author.

Conflicts of Interest: The authors declare no conflict of interest.

References

1. Erdinc, O.; Paterakis, N.G.; Mendes, T.D.P.; Bakirtzis, A.G.; Catalão, J.P.S. Smart Household Operation Considering Bi-Directional EV and ESS Utilization by Real-Time Pricing-Based DR. *IEEE Trans. Smart Grid* **2015**, *6*, 1281–1291. [CrossRef]
2. Tushar, W.; Yuen, C.; Huang, S.; Smith, D.B.; Poor, H.V. Cost Minimization of Charging Stations With Photovoltaics: An Approach With EV Classification. *IEEE Trans. Intell. Transp. Syst.* **2016**, *17*, 156–169. [CrossRef]
3. Shi, Y.; Tuan, H.D.; Savkin, A.V.; Duong, T.Q.; Poor, H.V. Model Predictive Control for Smart Grids With Multiple Electric-Vehicle Charging Stations. *IEEE Trans. Smart Grid* **2019**, *10*, 2127–2136. [CrossRef]
4. Wang, L.; Qin, Z.; Slangen, T.; Bauer, P.; van Wijk, T. Grid Impact of Electric Vehicle Fast Charging Stations: Trends, Standards, Issues and Mitigation Measures—An Overview. *IEEE Open J. Power Electron.* **2021**, *2*, 56–74. [CrossRef]
5. Pavić, I.; Pandžić, H.; Capuder, T. Electric vehicle based smart e-mobility system—Definition and comparison to the existing concept. *Appl. Energy* **2020**, *272*, 115153. [CrossRef]
6. Bellocchi, S.; Klöckner, K.; Manno, M.; Noussan, M.; Vellini, M. On the role of electric vehicles towards low-carbon energy systems: Italy and Germany in comparison. *Appl. Energy* **2019**, *255*, 113848. [CrossRef]
7. Jian, L.; Xue, H.; Xu, G.; Zhu, X.; Zhao, D.; Shao, Z.Y. Regulated Charging of Plug-in Hybrid Electric Vehicles for Minimizing Load Variance in Household Smart Microgrid. *IEEE Trans. Ind. Electron.* **2013**, *60*, 3218–3226. [CrossRef]
8. IRGC. *Demand-Side Flexibility for Energy Transitions: Ensuring the Competitive Development of Demand Response Options*; Technical Report; International Risk Governance Council (IRGC): Lausanne, Switzerland, 2015.
9. Palensky, P.; Dietrich, D. Demand Side Management: Demand Response, Intelligent Energy Systems, and Smart Loads. *IEEE Trans. Ind. Inform.* **2011**, *7*, 381–388. [CrossRef]
10. Haider, H.T.; See, O.H.; Elmenreich, W. A review of residential demand response of smart grid. *Renew. Sustain. Energy Rev.* **2016**, *59*, 166–178. [CrossRef]
11. Wang, Z.; Gu, C.; Li, F.; Bale, P.; Sun, H. Active Demand Response Using Shared Energy Storage for Household Energy Management. *IEEE Trans. Smart Grid* **2013**, *4*, 1888–1897. [CrossRef]
12. Asadinejad, A.; Tomsovic, K. Optimal use of incentive and price based demand response to reduce costs and price volatility. *Electr. Power Syst. Res.* **2017**, *144*, 215–223. [CrossRef]
13. Eid, C.; Koliou, E.; Valles, M.; Reneses, J.; Hakvoort, R. Time-based pricing and electricity demand response: Existing barriers and next steps. *Util. Policy* **2016**, *40*, 15–25. [CrossRef]
14. Kong, W.; Luo, F.; Jia, Y.; Dong, Z.Y.; Liu, J. Benefits of Home Energy Storage Utilization: An Australian Case Study of Demand Charge Practices in Residential Sector. *IEEE Trans. Smart Grid* **2021**, *12*, 3086–3096. [CrossRef]
15. Wang, L. *Model Predictive Control System Design and Implementation Using MATLAB*; Springer: London, UK, 2009.
16. Maciejowski, J.M. *Predictive Control with Constraints*; Prentice Hall: Harlow, UK, 2002.
17. Raimondi Cominesi, S.; Farina, M.; Giulioni, L.; Picasso, B.; Scattolini, R. A Two-Layer Stochastic Model Predictive Control Scheme for Microgrids. *IEEE Trans. Control. Syst. Technol.* **2018**, *26*, 1–13. [CrossRef]
18. Parisio, A.; Rikos, E.; Glielmo, L. A Model Predictive Control Approach to Microgrid Operation Optimization. *IEEE Trans. Control. Syst. Technol.* **2014**, *22*, 1813–1827. [CrossRef]

19. Qi, W.; Liu, J.; Christofides, P.D. Supervisory Predictive Control for Long-Term Scheduling of an Integrated Wind/Solar Energy Generation and Water Desalination System. *IEEE Trans. Control. Syst. Technol.* **2012**, *20*, 504–512. [CrossRef]
20. Hu, J.; Xu, Y.; Cheng, K.W.; Guerrero, J.M. A model predictive control strategy of PV-Battery microgrid under variable power generations and load conditions. *Appl. Energy* **2018**, *221*, 195–203. [CrossRef]
21. Lei, M.; Yang, Z.; Wang, Y.; Xu, H.; Meng, L.; Vasquez, J.C.; Guerrero, J.M. An MPC-Based ESS Control Method for PV Power Smoothing Applications. *IEEE Trans. Power Electron.* **2018**, *33*, 2136–2144. [CrossRef]
22. Shan, Y.; Hu, J.; Li, Z.; Guerrero, J.M. A Model Predictive Control for Renewable Energy Based AC Microgrids Without Any PID Regulators. *IEEE Trans. Power Electron.* **2018**, *33*, 9122–9126. [CrossRef]
23. Patrinos, P.; Trimboli, S.; Bemporad, A. Stochastic MPC for real-time market-based optimal power dispatch. In Proceedings of the 2011 50th IEEE Conference on Decision and Control and European Control Conference, Orlando, FL, USA, 12–15 December 2011; pp. 7111–7116.
24. Guo, Y.; Sheng, S.; Anglani, N.; Lehman, B. Optimal Power Management for Grid-Connected Microgrid Considering Modelling of Different Electricity Cost and Battery Degradation Cost. In Proceedings of the 2019 20th Workshop on Control and Modeling for Power Electronics (COMPEL), Toronto, ON, Canada, 16–19 June 2019; pp. 1–7.
25. Elkazaz, M.; Sumner, M.; Pholboon, S.; Davies, R.; Thomas, D. Performance Assessment of an Energy Management System for a Home Microgrid with PV Generation. *Energies* **2020**, *13*, 3436. [CrossRef]
26. Simmini, F.; Agostini, M.; Coppo, M.; Caldognetto, T.; Cervi, A.; Lain, F.; Carli, R.; Turri, R.; Tenti, P. Leveraging Demand Flexibility by Exploiting Prosumer Response to Price Signals in Microgrids. *Energies* **2020**, *13*, 3078. [CrossRef]
27. Tesla Powerwall. 2021. Available online: https://www.tesla.com/powerwall (accessed on 18 February 2021).
28. Rivera, S.; Kouro, S.; Vazquez, S.; Goetz, S.M.; Lizana, R.; Romero-Cadaval, E. Electric Vehicle Charging Infrastructure: From Grid to Battery. *IEEE Ind. Electron. Mag.* **2021**, *15*, 37–51. [CrossRef]
29. Plötz, P.; Jakobsson, N.; Sprei, F. On the distribution of individual daily driving distances. *Transp. Res. Part Methodol.* **2017**, *101*, 213–227. [CrossRef]
30. Torriti, J. Price-based demand side management: Assessing the impacts of time-of-use tariffs on residential electricity demand and peak shifting in Northern Italy. *Energy* **2012**, *44*, 576–583. [CrossRef]
31. ARERA. *DCO 322/2012/R/eel: Review of the Integrated text on "Scambio sul Posto" Measure [Revisione del Testo Integrato per lo Scambio sul Posto]*; ARERA: Milano, Italy, 2012.
32. Australian Government. Energymadeeasy. 2021. Available online: https://www.energymadeeasy.gov.au/ (accessed on 24 February 2021).

energies

Article

Electric Vehicle Fleets as Balancing Instrument in Micro-Grids

Giambattista Gruosso *,† and Fredy Orlando Ruiz †

Dipartimento di Elettronica, Informazione e Bioingegneria—DEIB, Politecnico di Milano, Piazza Leonardo da Vinci 32, 20133 Milano, Italy; fredy.ruiz@polimi.it
* Correspondence: giambattista.gruosso@polimi.it; Tel.: +39-02-2399-3696
† These authors contributed equally to this work.

Abstract: Micro-grids have become the building block of modern energy systems, where distributed resources are the characterizing feature. The charging operation of electric vehicles can be exploited as a flexible load to achieve operational goals of the micro-grid. In the particular case of car-sharing fleets, the degrees of freedom in the charging procedures are reduced when compared to private users. In this work, we illustrate how a car sharing fleet can be incorporated as a flexible load in the micro-grid management system. A linear optimization problem is formulated, where the cost function makes a trade-off between the gain in flexibility in the micro-grid and the loss incurred by the car-sharing service for delaying the recharging procedure of the EV. The proposed approach is evaluated on a data set of charging events generated by a real car-sharing fleet showing that the EMS allows reducing the daily peak demand requested to the public grid and diminishes the operational costs.

Keywords: micro-grid planning; electrical vehicles; energy storage; flexible programming

Citation: Gruosso, G.; Ruiz, F.O. Electric Vehicle Fleets as Balancing Instrument in Micro-Grids. *Energies* **2021**, *14*, 7616. https://doi.org/10.3390/en14227616

Academic Editors: Maria Carmela Di Piazza and Javier Contreras

Received: 7 October 2021
Accepted: 11 November 2021
Published: 15 November 2021

1. Introduction

Today, nano and micro-grids play a strategic role in the development of the electricity system [1,2]. They represent, most of the time, autonomous energy communities [3], able to guarantee regulated energy exchanges, reducing the stress on the public distribution network. In these networks, it is possible to integrate different energy sources and at the same time new and programmable loads could be introduced. It becomes crucial to correctly plan energy flows in this context to optimize efficiency [4–7]. micro-grid architectures have different configurations, but in most cases the high use of renewable sources, such as solar, make these networks, on the one hand to behave with low inertia and on the other hand to count on generation resources, whose actual production is not easily predictable compared to the needs of the loads. It is, therefore, essential to have energy storage systems capable of promoting an optimal use of the resources. One of the most interesting and challenging solutions is to use Electric Vehicles (EV) as virtual storage systems [8–11]. Different optimization-based solutions to manage the charging profile of electric vehicles were proposed [12]. Most of the works assume that the electric vehicles are autonomous and an important effort is made in forecasting the arrival time of each EV and other uncertain variables, such as the arrival State of Charge (SoC) [13,14], or in developing robust scheduling techniques that can guarantee an adequate power flow to/from the public grid, in front of any feasible realization of arrival time and *SoC* [15].

The existing solutions to manage the optimal charging of EV fleets have two main limitations. First, most of them need to track the dynamics of each vehicle in the fleet to accurately estimate its arrival time and *SoC*. This approach poses a strong limitation on the scalability of the solution, as the complexity of the estimation and optimization problems grows linearly with the size of the fleet. In [12], a different approach is presented, where the main element of the model is the charging station and the arrival of EVs are handled as events that modify the state of the station. This configuration, limits the size of the problems to the number of stations handled by the management system. On the other hand, in the literature, the degrees of freedom that are considered to manipulate the charging profile of EVs are the power level and the charging interval, given some constraints on the total recharging time. However, none of the existing models takes into account the cost incurred by the fleet manager with a longer charging time. In the specific case of a car-sharing fleet, the operator of the sharing service actually suffers a loss when a EV is unavailable during the recharge event. The longer the charging time, the higher the loss.

In this paper, we present a method for planning a micro-grid operation in which renewable sources, electrical loads and a car-sharing fleet of EVs are present. The vehicles energy requests are extracted from a real use case of a car-sharing fleet. The proposed solution takes into account the operational policy of the car-sharing company to gain information about the structure of the problem, impose costs and constraints on the operation, and reduce uncertainty in the energy demand. An optimal control problem is formulated, considering that all the EVs are taken to recharge mode at the same state of charge level. Flexibility is added to the recharging process augmenting the interval available to achieve the target *SoC* status, at the cost of reducing the time available for the car-sharing service. A multi-objective cost function allows performing a trade-off between flexibility in the demand curve and effectiveness of the sharing service. The validity of the formulation is verified in different scenarios of load and renewable generation.

In Section 2, we present the problem setup including the description of the employed data bases of EV charging events, loads and local photo-voltaic generation. Section 3 describes the operation of the car-sharing service and the resulting constraints and costs for the micro-grid operation. Section 4 describes the proposed energy management system, the considered multi-objective optimization problem and constraints. Section 5 illustrates the characteristics of the solution through a numerical example. Finally, Section 6 gives some conclusions and future developments .

2. Problem Setup

The considered micro-grid is formed by an industrial and commercial site, constituted by two circuits, a photo-voltaic (PV) generation plant and a set of charging stations that serve a fleet of shared electric vehicles.The grid is modeled as a single bus with limited capacity to exchange power with the main grid as shown in Figure 1.

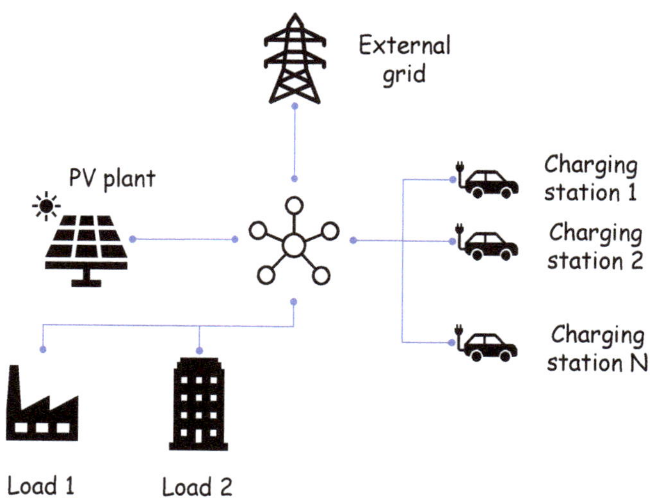

Figure 1. Micro-grid structure.

The data used to model the behaviour of the EV fleet are those collected in the city of Milan in Italy during the Teinvein project [16]. This project monitored a fleet of electric vehicles used for car-sharing. The available data set contains recharging events performed in the around 200 public charging stations placed in the urban area.

By analyzing the behavior of the charging stations, it is possible to derive usage profiles that take into account the amount of energy withdrawn and the initial state of charge of the vehicles [17]. Using a fleet of vehicles, as opposed to individual owners data, allows for realistic estimations of virtual battery behavior. The fleet manager can plan how many cars to allocate to virtual storage and how many to make available to users. In addition, the fleet manager can define charging strategies. For example, in the considered case study, recharging occurs when the state of charge of the vehicles is around 20%.

The arrival of the vehicles is asynchronous and can happen at very different moments. Moreover, each car can leave the location only when it has reached a predefined target *SoC* named $\overline{SoC}$. Thus, the scheduling problem we set has to deal with challenging constraints to satisfy. Starting from the methodology proposed in [17] it is possible to obtain the occupation curves of the charging stations, such as for example those shown in Figure 2, where daily profiles of charging station occupation is shown for a set of 10 consecutive days. Each color represents a different day. It can be noticed that for this station the maximum occupation is 4 simultaneous EVs connected for recharging, and that several days present intervals without any EV connected to the station, thus limiting the flexibility of the micro-grid.

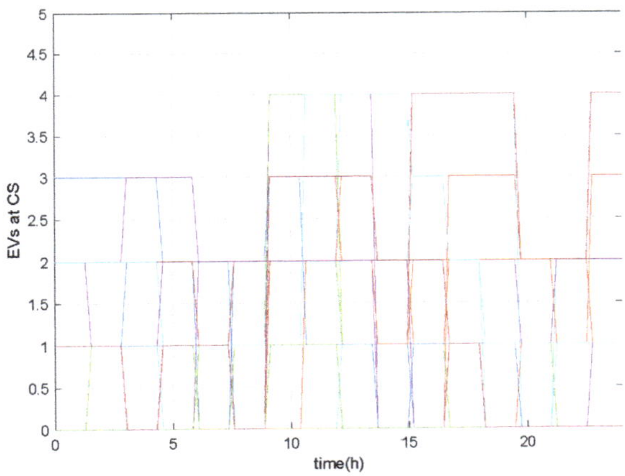

Figure 2. Sample of a Charging station occupation for one month.

As far as the modelling of loads and photo-voltaic sources is concerned, we rely on those presented in [18]. In that paper a multi-year database containing data from a real micro-grid is presented. The loads of the two circuits are assumed partially flexible and their typical behaviour is shown in Figure 3, where the daily profile of the load is illustrated for one year. Each color represents a different day. It can be noticed that the peak load occurs between the hours 10 and 18, also that the daily baseline can vary from 40 kW up to 70 kW.

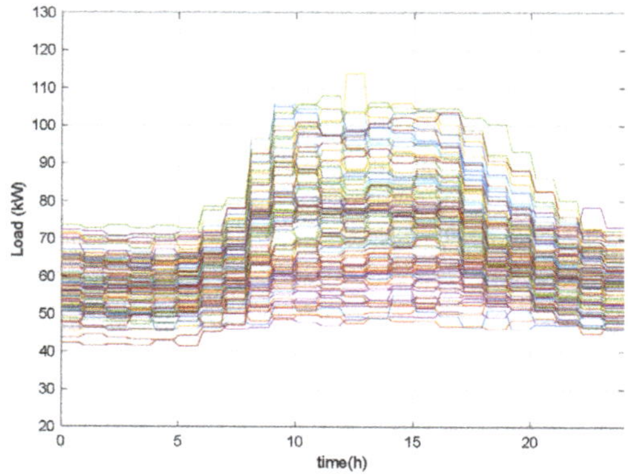

Figure 3. Sample of hourly demand profiles in the data set.

Similarly, photo-voltaic generation profiles are extracted form the data-base presented in [18]. In Figure 4, the daily generation profiles are reported for one year. Each color represents a different day. Please note that the daily peak production can rise up to 83 kW.

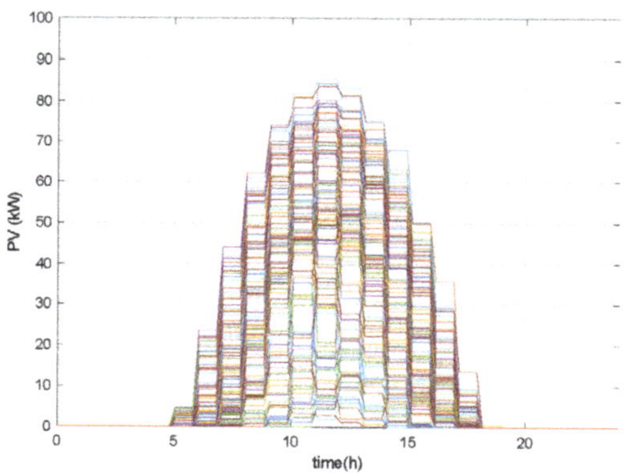

Figure 4. Sample of hourly PV plant generation profiles in the data set.

Table 1 shows the main parameters of the considered micro-grid.

Table 1. Microgird parameters.

Element	Maximum Power (kW)	Average Power (kW)
Load 1	120	68
Load 2	120	84
PV plant	90	10
External grid	250	N.D.
Charging station (each)	3	N.D.

Finally, the cost of the energy acquired from the public grid, considered in the case study, is a standard time-of-use rate with three different price ranges:

- early morning/late evening
- mid-day
- peak-hours

Management System

The energy management systems is the gateway between the controllers of the main elements of the micro-grid and the public grid. It uses the forecasts on the arrival time of vehicles, and the trend of loads and renewable sources to make optimal decisions about flexible loads activation and EV charging profiles. The architecture is shown in Figure 5. Here, the lines represent information flows and the arrows their direction. That means that the EMS obtains information from the forecasts providers about generation, EV charging events, load profiles and energy tariffs, but without affecting their decisions. On the other hand, the EMS interacts with flexible loads and charging stations in a two-way communication. In one direction it obtains information about the real-time status of the consumption, charging station status, connected EVs, etc., and, on the other, it commands how to activate flexible loads and how to modify the charging profile of each EV connected to the station.

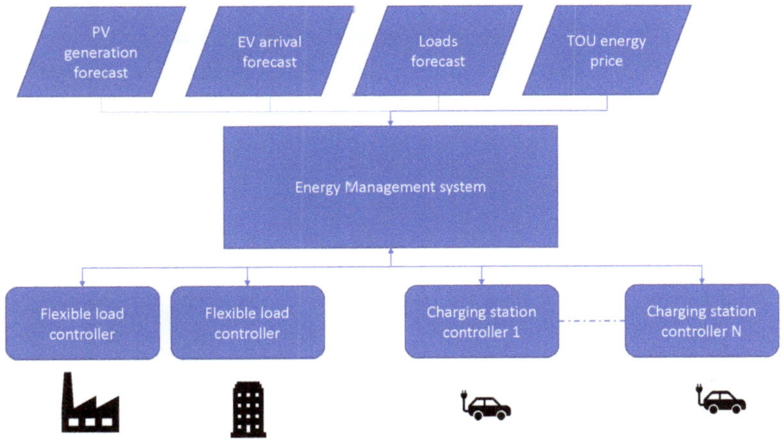

Figure 5. Structure of the Energy Management System.

The management system works to make sure that the power balance of the micro-grid is satisfied at each instant, while optimizing a set of Key Performance Indicators.

The power balance of the system is:

$$P_{grid}(t) + P_{PV}(t) = P_{L1}(t) + P_{L2}(t) + \sum_{i=1}^{N} P_i^{CS}(t) \tag{1}$$

where $P_{grid}(t)$ is the power absorbed to the main grid, $P_{PV}(t)$ is the power generated by the photo-voltaic plant, $= P_{Li}(t)$ is the power consumed by each circuit load, and $P_i^{CS}(t)$ is the power consumed by each charging station.

Please note that if the loads are inflexible and the charging stations operate without any management system, recharging each EV at maximum power from its arrival until achieving the target *SoC* condition, the power balance can only be guaranteed by modifying the power absorbed to the main grid.

The aim of the energy management system is to offer a feasible power flow while maintaining each component within its technical limits, exploit the flexibility of the EVs charging process and of the loads, to optimize the operational cost of the grid and minimizing the power exchanges with the main grid.

To provide some flexibility to the micro-grid, in the literature it was proposed to install local energy resources, such as traditional generators and/or static storage systems. However, if none of these resources are available, adding a management system to the charging stations can allow the micro-grid to achieve the proposed objectives at the expense of increasing the charging time of the electric vehicles.

3. EV Fleet Recharging Policy

As shown in [17], in the present framework we are dealing wit a set of uniform vehicles with the same decision criteria, all the EVs are taken to the charging station when the *SoC* achieves a prescribed level and all of them are given the same time window to come back to operation. The aim of this section is to formulate a formal description of the recharging policy, to be employed later in the optimization problem solved by the EMS.

As described at the beginning of Section 2, the operation policy of the charging station entails the continuous supervision of the *SoC* of each vehicle and the decision to recharge the vehicle whenever its state goes below a prescribed level. Let $\sigma_i(t)$ be a binary variable representing the operation mode of vehicle i, where $\sigma_i(t) = 0$ means that vehicle i is operating in car-sharing mode during the time interval t and $\sigma_i(t) = 1$ means that at period t the vehicle is in recharging mode.

The policy can be summarized as follows:

1. While an EV is in car sharing service, supervise its *SoC* and
$$if\ \ SoC_i(t) \geq \underline{SoC}\ \ and\ \ \sigma_i(t) = 0\ \ \Rightarrow \sigma_i(t+1) = 0; \tag{2}$$
2. When the *SoC* of the EV drops below $\underline{SoC}$ end the car-sharing operation and bring it to the charging station.
$$if\ \ SoC_i(t) < \underline{SoC}\ \ and\ \ \sigma_i(t) = 0\ \ \Rightarrow \sigma_i(t+1) = 1\ \ and\ \ t_{ai} = t. \tag{3}$$
3. At the beginning of the charging process set the target departure time from the charging station as
$$t_{di} = t_{ai} + \Delta_T \tag{4}$$
where Δ_T is the maximum allowed recharging time.
4. At the charging station, supervise the recharging procedure and bring back the EV to service as soon as the *SoC* rises above $\overline{SoC}$:
$$if\ \ SoC_i(t) \geq \overline{SoC}\ \ and\ \ \sigma_i(t) = 1\ \ \Rightarrow\ \ \sigma_i(t+1) = 0. \tag{5}$$

Introducing $X_i(t)$ as the state variable representing the state of charge of charging station i, the previous recharging policy leads to the following dynamics from the point of view of the charging station:

$$X_i(t+1) = \begin{cases} 0 & \text{if } \sigma_i(t) = 0 \\ SoC_o & \text{if } \sigma_i(t) = 1 \text{ and } \sigma_i(t-1) = 0 \\ X_i(t) + \delta_t \eta \sigma_i(t) P_i^{CS}(t) & \text{if } \sigma_i(t) = 1 \text{ and } \sigma_i(t-1) = 1 \end{cases} \tag{6}$$

Please note that the previous formulation decouples the usage of each EV and the recharging process.

4. Flexibility-Based Energy Management System

In this section, we describe the proposed a novel EMS that relies on the particular recharging policy of the EV fleet.

The EMS is based on an multi-objective optimization problem with uncertain constraints.

The micro-grid incurs in different costs during normal operation. The main cost is the provision of energy from the main grid. Considering a Time of Use (ToU) tariff the grid energy cost for a given time interval is:

$$J_{grid} = \sum_{k=0}^{T_H} \pi_{grid}(t+k) P_{grid}(t+k)\delta_t \tag{7}$$

where $\pi_{grid}(t)$ is the cost per kWh of energy bought to the main grid.

At each charging station, there is a cost associated with the time elapsed for the charging process of each vehicle. It is defined as an increasing function of the recharging interval whenever the process takes longer than the minimum feasible charging time, that is,

$$J_{CSi} = \sum_{k=0}^{T_H} step(k - \Delta_C)\lfloor \overline{SoC} - SoC(t+k) \rfloor \tag{8}$$

where Δ_C is the shortest interval required to achieve the target SoC when the charging station applies the maximum acceptable power to the EV. Figure 6 illustrates the time evolution of the cost J_{CSi}.

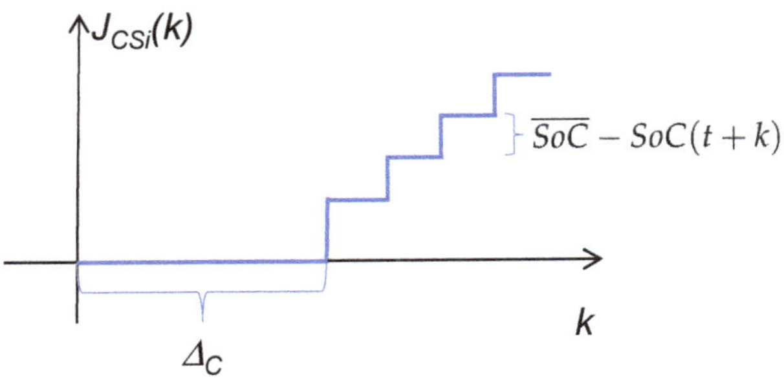

Figure 6. Time evolution of the cost associated with the delayed charging process.

For each circuit of the micro-grid, it is assumed that there is a set of referable loads that can provide flexibility to the operation. The activation of these loads has a cost related to the discomfort caused to the users of the services plus the operational cost of the demand management systems. The capacity of the flexibility service is limited to 10% of the forecast power at each interval, while the daily consumption must keep constant. The costs related to the flexible loads can be expressed as:

$$J_{FLi} = \sum_{k=0}^{T_H} \pi_{FL}\delta_t \lfloor \widehat{P}_{Li}(t) - P_{Li}(t) \rfloor \tag{9}$$

where π_{FL} is the reduction cost per kWh and $\widehat{P}_{Li}(t)$ is the forecast consumption at time t.

Finally, The effective consumption at time t is constraint as:

$$0.9\widehat{P}_{Li}(t) \leq P_{Li}(t) \leq 1.1\widehat{P}_{Li}(t) \tag{10}$$

and the daily consumption as:

$$\sum_{t=0}^{T_{end}} \widehat{P}_{Li}(t) = \sum_{t=0}^{T_{end}} P_{Li}(t). \tag{11}$$

The problem of the EMS is to minimize the different costs related to the micro-grid operation, manipulating the power applied to the EVs during the recharge interval and activating flexible loads when necessary.

For a receding horizon of length N, the problem faced by the EMS can be stated as:

$$\min_{(P_i^{CS}(k),P_{Li}(k)),\, k=t,\, t+1,\, \dots,\, t+T_H} J_{grid} + J_{FL1} + J_{FL2} + \sum_{i=1}^{N} J_{CSi} \tag{12}$$

Subject to

$$P_{grid}(k) = \widehat{P}_{L1}(k) + \widehat{P}_{L2}(k) + \sum_{i=1}^{N} P_i^{CS}(k) - \widehat{P}_{PV}(k);\ k = (t, t+1, \dots, T_H)$$

Car-sharing policy and charging station dynamics, Equations (2)–(6)

Flexible loads constraints, Equations (10) and (11)

where $\widehat{P}_{L1}(k)$ and $\widehat{P}_{L2}(k)$ are the estimates of the load power for interval k and $\widehat{P}_{FV}$ is the PV power generation estimate at time k.

Please note that to solve problem (12), the EMS must have access, as well as to the forecast of load and PV generation, also to the arrival and departure information of the EVs for each charging station, i.e., signals $\sigma(t)$ and SoC_0. These signals are not know a priori, but the recharging policy of the fleet and historical data of the fleet operation allows building adequate estimators. For example, the arrival SoC of each vehicle is assumed to be $\underline{SoC}$, as this is the threshold to start a charging event. About the arrival time, each time the optimization algorithm is executed, an average number of charging events is estimated for the prediction horizon T_H, based on the data base of charging events. In this form, the optimal management of the currently connected vehicles is obtained for the most likely scenario. Then, at the next sampling time, the actual number of connected vehicles, the power forecasts and the expected EV arrivals are updated, generating a new power management policy in a receding horizon setup.

About the complexity of problem (12), it is a linear program, whose dimensionality depends on the number of flexible loads circuits and charging stations, but not on the size of the EV fleet. It can be noticed that all the terms in the cost function depend linearly on the decision variables, the power balance is a linear constraint, flexible loads capacity and daily limits are also linear constraints, while the car-sharing policy and charging station dynamics involve also linear equality and inequality constraints. Please note that from the point of view of the optimizer, the status of each charging station $\sigma_i(t)$ is a known binary signal, not a decision variable. These characteristics make the algorithm efficient ant scalable.

5. Numerical Results

The EMS proposed in the previous section was evaluated on a 235 days long simulation. The load and PV generation data are extracted from the data base reported in [18]. The EV fleet recharging events are taken from the data base reported in [17].

The sampling time of the EMS, δ_t is selected as 0.25 h. This interval is fast enough to update the scheduling whenever a EV is connected to a charging station. The minimum recharging interval Δ_C is set as 4 hours, that is the period required to fully charge the vehicle using the highest power level, without any flexibility consideration. The prediction horizon T_H is fixed as 8 h. This length allows planning the complete recharging process of all the EVs connected to the CSs at a given sampling instant.

Figure 7 shows a box-plot of the peak power requested to the grid. It can be seen that the proposed EMS reduces both the worst-case power request to the grid and the mean of the peak power along the 235 days. The worst-case peak power is reduced from 123 kW to 112 kW and the median from 105 kW to 99 kW . this effect can alleviate congestion events in the grid. Figure 8 shows a histogram of the daily energy cost of the micro-grid. For each price bin, the vertical axis reports the percentage of the 235 simulated days that has a cost within the bin interval. It is observed that the operational costs are reduced by 3% in this simulation. Finally, Figure 9 shows a sample of the daily load profile of the micro-grid. Please note that the EMS reduces the peak and generates longer charging intervals. This modification can cause a loss to the fleet operator. However, the savings in energy provision can compensate the longer recharging times.

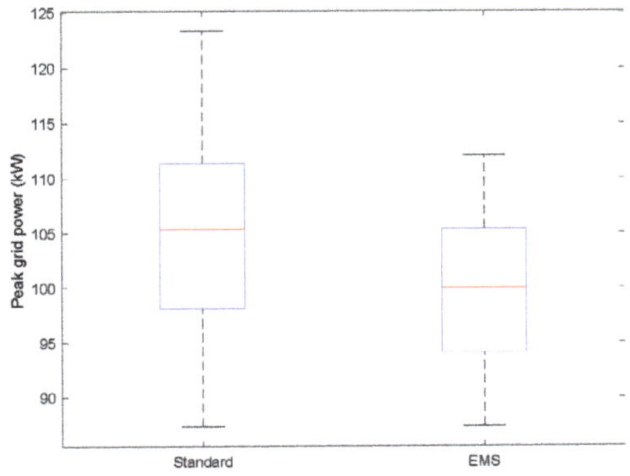

Figure 7. Boxplot of peak power requested to the main grid under standard (**left**) and optimal (**right**) EMS for the EV fleet.

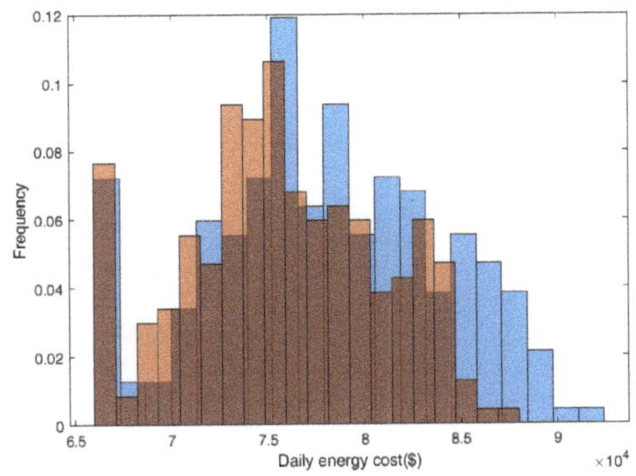

Figure 8. Histogram of daily energy cost of the micro-grid under standard (blue) and optimal (orange) EMS for the EV fleet.

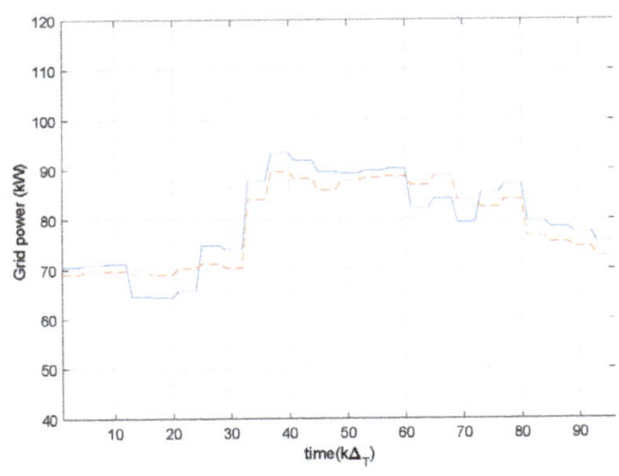

Figure 9. Sample daily profile of power requested to the main grid under standard (blue) and optima (orange) EMS for the EV fleet.

6. Conclusions

In this work, we proposed an energy management system for a micro-grid that exploits the particularities of an EV fleet used in a car-sharing service. The system schedules the charging power of each vehicle considering the energy cost and forecast information about load and renewable generation. The cost function allows making a trade-off between flexibility exploitation to reduce energy costs and the lost in the car-sharing service caused by a longer recharging time. The methodology is evaluated through a long simulation campaign, considering experimental data of load profiles, PV generation and EV charging events from a car-sharing fleet. The results show that the EMS is able to reduce the peak power consumption of the micro-grid, extending the charging time from 4 to 5 or 6 h. Further investigation is required to evaluate the sensitivity of the solution to uncertainty in the forecast information about fleet behavior, load and PV generation.

Author Contributions: Conceptualization, G.G. and F.O.R.; methodology, G.G. and F.O.R.; software, F.O.R.; validation, G.G. and F.O.R.; data curation, G.G. and F.O.R.; writing—original draft preparation, G.G. and F.O.R.; writing—review and editing, G.G. and F.O.R. All authors have read and agreed to the published version of the manuscript.

Funding: This research received no external funding.

Institutional Review Board Statement: Not applicable.

Data Availability Statement: No data are available.

Conflicts of Interest: The authors declare no conflict of interest.

References

1. Moreno-Munoz, A. Special Issue "Nanogrids, Microgrids, and the Internet of Things (IoT): Towards the Digital Energy Network". *Energies* **2019**, *12*, 3878. [CrossRef]
2. Gandhi, K.; Gupta, S. Operational strategies and electricity market structure of Microgrid: A Critical Review. *Renew. Energy Focus* **2021**, *39*, 163–171. [CrossRef]
3. Perez-DeLaMora, D.; Quiroz-Ibarra, J.E.; Fernandez-Anaya, G.; Hernandez-Martinez, E. Roadmap on community-based microgrids deployment: An extensive review. *Energy Rep.* **2021**, *7*, 2883–2898. [CrossRef]
4. Bagheri Tookanlou, M.; Marzband, M.; Al Sumaiti, A.; Mazza, A. Cost-benefit analysis for multiple agents considering an electric vehicle charging/discharging strategy and grid integration. In Proceedings of the 2020 IEEE 20th Mediterranean Electrotechnical Conference (MELECON), Palermo, Italy, 16–18 June 2020; pp. 19–24. [CrossRef]
5. Al-Ismail, F.S. DC Microgrid Planning, Operation, and Control: A Comprehensive Review. *IEEE Access* **2021**, *9*, 36154–36172. [CrossRef]
6. Rafiee Sandgani, M.; Sirouspour, S. Energy Management in a Network of Grid-Connected Microgrids/Nanogrids Using Compromise Programming. *IEEE Trans. Smart Grid* **2018**, *9*, 2180–2191. [CrossRef]
7. Lee, S.; Jin, H.; Vecchietti, L.F.; Hong, J.; Har, D. Short-Term Predictive Power Management of PV-Powered Nanogrids. *IEEE Access* **2020**, *8*, 147839–147857. [CrossRef]
8. Dang, Q.; Wu, D.; Boulet, B. EV Fleet as Virtual Battery Resource for Community Microgrid Energy Storage Planning Le parc de véhicules électriques comme ressource de batterie virtuelle pour la planification du stockage d'énergie des micro-réseaux communautaires. *IEEE Can. J. Electr. Comput. Eng.* **2021**, 1–12. [CrossRef]
9. Vuelvas, J.; Ruiz, F.; Gruosso, G. A time-of-use pricing strategy for managing electric vehicle clusters. *Sustain. Energy Grids Netw.* **2021**, *25*, 100411. [CrossRef]
10. Dinkhah, S.; Negri, C.A.; He, M.; Bayne, S.B. V2G for Reliable Microgrid Operations: Voltage/Frequency Regulation with Virtual Inertia Emulation. In Proceedings of the 2019 IEEE Transportation Electrification Conference and Expo (ITEC), Detroit, MI, USA, 19–21 June 2019; pp. 1–6. [CrossRef]
11. Vuelvas, J.; Ruiz, F.; Gruosso, G. Energy price forecasting for optimal managing of electric vehicle fleet. *IET Electr. Syst. Transp.* **2020**, *10*, 401–408. [CrossRef]
12. Diaz-Londono, C.; Colangelo, L.; Ruiz, F.; Patino, D.; Novara, C.; Chicco, G. Optimal Strategy to Exploit the Flexibility of an Electric Vehicle Charging Station. *Energies* **2019**, *12*, 3834. [CrossRef]
13. Diaz-Londono, C.; Ruiz, F.; Mazza, A.; Chicco, G. Optimal Operation Strategy for Electric Vehicles Charging Stations with Renewable Energy Integration. *IFAC-PapersOnLine* **2020**, *53*, 12739–12744. [CrossRef]
14. Giordano, F.; Arrigo, F.; Diaz-Londono, C.; Spertino, F.; Ruiz, F. Forecast-Based V2G Aggregation Model for Day-Ahead and Real-Time Operations. In Proceedings of the 2020 IEEE Power Energy Society Innovative Smart Grid Technologies Conference (ISGT), Washington, DC, USA, 17–20 February2020; pp. 1–5. [CrossRef]

15. Badami, M.; Fambri, G.; Mancò, S.; Martino, M.; Damousis, I.G.; Agtzidis, D.; Tzovaras, D. A Decision Support System Tool to Manage the Flexibility in Renewable Energy-Based Power Systems. *Energies* **2020**, *13*, 153. [CrossRef]
16. Gajani, G.; Bascetta, L.; Gruosso, G. Data-driven approach to model electrical vehicle charging profile for simulation of grid integration scenarios. *IET Electr. Syst. Transp.* **2019**, *9*, 168–175. [CrossRef]
17. Gruosso, G.; Mion, A.; Storti Gajani, G. Forecasting of electrical vehicle impact on infrastructure: Markov chains model of charging stations occupation. *eTransportation* **2020**, *6*, 100083. [CrossRef]
18. Vink, K.; Ankyu, E.; Koyama, M. Multiyear microgrid data from a research building in Tsukuba, Japan. *Sci. Data* **2019**, *6*, 190020. [CrossRef] [PubMed]

MDPI
St. Alban-Anlage 66
4052 Basel
Switzerland
Tel. +41 61 683 77 34
Fax +41 61 302 89 18
www.mdpi.com

Energies Editorial Office
E-mail: energies@mdpi.com
www.mdpi.com/journal/energies

www.ingramcontent.com/pod-product-compliance
Lightning Source LLC
Chambersburg PA
CBHW062047220926
51850CB00010B/2690
9783036525587